D1020042

TERRORS AND MARVELS

BY TOM SHACHTMAN

Terrors and Marvels

"I Seek My Brethren"

Absolute Zero and the Conquest of Cold

Around the Block

The Most Beautiful Villages of New England

The Inarticulate Society

Skyscraper Dreams

Decade of Shocks, 1963–1974

The Phony War, 1939–1940

Edith and Woodrow

The Day America Crashed

COLLABORATIVE BOOKS

Torpedoed (with Edmond D. Pope)

"I Have Lived in the Monster" (with Robert K. Ressler)

Justice Is Served (with Robert K. Ressler)

Whoever Fights Monsters (with Robert K. Ressler)

Image by Design (with Clive Chajet)

Straight to the Top (with Paul G. Stern)

The Gilded Leaf (with Patrick Reynolds)

The FBI-KGB War (with Robert Lamphere)

FOR CHILDREN

The President Builds a House

Video Power (with Harriet Shelare)

America's Birthday

Parade!

The Birdman of St. Petersburg

Growing Up Masai

NOVELS

Beachmaster

Wavebender

Driftwhistler

TERRORS AND MARVELS

How Science and Technology Changed the Character and Outcome of World War II

TOM SHACHTMAN

wm

WILLIAM MORROW

An Imprint of HarperCollins*Publishers*

FIRST EDITION

Designed by Paula Russell Szafranski

Printed on acid-free paper

Library of Congress Cataloging-in-Publication Data has been applied for.

ISBN 0-380-97876-8

02 03 04 05 06 WBC/QW 10 9 8 7 6 5 4 3 2 1

To Harriet:

Twenty-five and counting . . .

Contents

Terror from the Sky

IN THE SKIES above London, electronically sophisticated, computer-controlled, robotic mechanical destroyers clash. Below, emergency crews race to put out firestorms and rescue the victims of the incoming robo-bombs, while the rest of the populace watches the battle in awe and awaits the outcome, which can decide their fate.

This scene could be a twenty-first century nightmare, a depiction of the electronic battlefield of the future. But it took place on June 13, 1944.

It was one week to the day after the Allies invaded Normandy in an attempt to end the domination of Europe by the Third Reich. Now Adolf Hitler was striking back, as Londoners recalled that he had threatened to do. For several years, Hitler had warned his enemies that if the Reich was attacked, he would retaliate with weapons that would wreak great damage, terrorize Germany's enemies and compel them to their knees—marvels never seen before, the products of German scientific expertise that he had been developing and accumulating toward such a moment.

Hitler's jet-powered flying bombs, called V-1s, aimed at the tallest buildings and most cherished landmarks, carried explosive payloads capable of flattening several office buildings at once. Hundreds of them each day hurtled across the English Channel from camouflaged ski-ramp launchers on the French and Belgian coasts. So fast did the unmanned bombs travel that they materialized in the air above their targets only moments before impact, screeching and flaming across the sky, then striking a building and exploding with devastating and horrifying results. To Londoners in 1944, these terror weapons seemed straight out of the science-fiction novels of H. G. Wells, or reminiscent of the magical wizardry of Merlin in King Arthur's court.

The machines were all the more awful because they were unexpected; the government had not warned ordinary Britons that such bombs might be coming (even though the War Cabinet had known of the possible terror for a year). The first day, one bomb landed at Swanscombe, another at Bethnal, luckily killing only six people. The next day a hundred more rained down, many making direct hits on palaces, government buildings, and financial offices and coming so swiftly that Londoners had no time to reach air-raid shelters before the bombs exploded nearby. Dozens of people died, whole blocks of flats and offices collapsed.

A robot bomb that did not burst on landing was carted off by the military for analysis. Britons wondered: Where was the response from the Allied wizards? Didn't the Americans have some secret weapon with which to counter the robotic bombs? Was there nothing the Allies could do to defend against these products of the Nazi geniuses?

Prime Minister Winston Churchill and his cabinet were frightened that the next batch of robot bombs might spread poison gas, or biological agents such as plague, or even—the worst nightmare—

contain an explosive device that made use of the power derived from the recently discovered splitting of the atomic nucleus of uranium. The Allied leaders knew, even if the public was unaware, that Germany had been trying to create various doomsday machines. Even when none of these nightmares was realized, the British government decided to evacuate a million Londoners.

"London is chaotic with panic and terror. The roads are choked with fleeing refugees," Reich minister of propaganda Joseph Goebbels gloated over German airwaves. The Nazis had learned from British radio broadcasts that the British government was acknowledging the damage caused by the V-1s—5,000 deaths, 15,000 injuries, hundreds of firestorms. The response of the German-speaking public to Goebbels' announcement was ecstatic: With this strange weapon, the V-1, they now had reason to hope Hitler might yet snatch victory for Germany from the jaws of defeat.

The Allied military command worried that shortly something worse would be coming from Germany: huge rockets known as V-2s, many stories high, with payloads twice as large and potent as those carried by the V-1s—weapons against which the Allies could not even imagine a defense. Beyond the V-2, the Allies feared, the Germans were readying the V-3, the *Hochdruckpumpe,* an enormous gun over 400 feet in length, a series of barrel chambers with sections protruding from their sides like ribs; by means of successive explosions in the side chambers, a 300-pound shell could be speeded up while passing through the main chamber until it emerged from the barrel at a tremendous velocity, easily able to reach London from France in a matter of minutes, too quickly for anything to prevent it from finding its target. Fifty V-3s were being assembled and aimed at industrial sites and at London.

To stave off the mounting terror that could potentially force Great Britain to sue for peace, the Allies now launched robots of their own, whose task it was to find the V-1s in the sky, track them through any twists and turns in flight, and destroy them before they could hit their targets. The Allied robots were even more complex than the V-1s, consisting of three mechanical marvels: two of them made up a robotic gun on the ground that sensed the approach of the V-1 when it could not even be seen by the human eye, and fired off the third, a projectile that when in the air acted as though it had a mind of its own by homing in relentlessly and inescapably on the V-1 and exploding itself and the attacking robot to smithereens. That was the theory, but although the Allies had been developing the underlying sensing technology of microwave radar for three years, the proximity fuze and the radio-wave-based artillery-fire director that controlled their robot system's destructive force had been more hastily developed in laboratories spread out across Great Britain, the United States, and Canada—and had not yet been tested together in battle.

Within days of the first V-1 attack, these untested Allied robots joined the battle of the flying unmanned destroyers. It was an epic clash: two sets of robotic warriors, each reflecting the work and imagination of their side's scientists and technologists, and beyond that, the social hierarchies and moral imperatives of the warring nations. And at the moment of the robots' meeting in the skies above Great Britain, no one alive could predict with certainty which belligerent group's mechanical devices would prevail—and in so doing, would assure one side's survival and hasten the other's demise.

Prologue: Legacies of the Great War

AMONG THE MILLIONS of deaths in the Great War, none was more senseless than that of the twenty-seven-year-old atomic physicist Henry Gwyn Jeffreys Moseley. Though serving in an engineer's battalion and not as a front-line soldier, he was killed at Gallipoli in 1915, where the British suffered 200,000 casualties while failing to accomplish their military objective. Moseley had been a rising star, expected one day to earn a Nobel Prize. He had already articulated "Moseley's Law," which described the frequency with which X rays were emitted from an element during bombardment of its nucleus, a measurement that would later be important in achieving nuclear fission.

The notion that science was above or beyond politics and war was echoed in a German scientific magazine's obituary of Moseley, printed while Germany and Great Britain were battlefield enemies. And in *Nature,* Ernest Rutherford, the leader of the Cavendish Laboratory at Cambridge, in which Moseley had worked, mourned his colleague's death and drew from it a larger lesson regarding the interpenetration of science, politics, and government:

> It is a national tragedy that our military organization . . . was so inelastic as to be unable, with a few exceptions, to utilize the offers of service of our scientific men except as combatants. . . . We cannot but recognize that [Moseley's] services would have been far more useful to his country in one of the numerous fields of scientific inquiry rendered necessary by the war than by exposure to the chances of a Turkish bullet.

In the 1930s Moseley's death would be cited as a prime example of the need to make a register of scientists and their specialties, so that if another war broke out, scientists could be commandeered to work in laboratories in support of the war effort, rather than in combat.

It was not as though there had been no such scientific contributions to the Great War. There had been. On the German side they were exemplified by chemist Fritz Haber's perfecting of a technique for fixing nitrogen from the air for use in gunpowder, which prolonged the war by helping Germany survive a blockade that had cut off its overseas supplies of nitrates; and by Haber's concocting of poison gases, which escalated the gas warfare that produced 100,000 deaths and 1.3 million injuries. Poison gas was the first scientific mass terror weapon, and for that, at the war's end, some British and French officials sought to charge Haber as a war criminal. But the effort to hold him responsible soon faded, and fellow scientists acknowledged the marvel of the nitrogen synthesis—and their independence of political considerations—by awarding Haber the Nobel Prize.

On the side of the Allied Powers, important scientifically based warfare tools were constructed by leading British scientists such as Rutherford and physicists William H. Bragg and his son William L. Bragg, joint winners of the 1915 Nobel Prize for their work on X-ray diffraction and crystal structures. They oversaw the development of sonar, which detected the presence of submarines by reflected sound pulses, and of the magnetic mine, which used a passing ship's iron to trigger an explosion. As science historian Roy MacLeod suggests, virtually every scientific discipline on both sides was tapped for use in the Great War:

> Geologists laid trench systems and tunnels, biologists designed camouflage, geographers drew artillery maps, physicists contrived acoustic devices, surgeons applied triage, psychologists discovered shell shock, and . . . bacteriologists and entomologists struggled to prevent the spread of infectious diseases. Darkly through the carnage, the "New Warfare," dominated by artillery, machine guns, wireless, submarines, and the internal combustion engine, became a surreal spectacle—as one French scientist put it, a "grandiose physical phenomenon"—a huge laboratory of deadly trial and error.

The horror of that war would remain fresh in the scientists' minds, emerging during the interwar years as an almost universally shared pacifism that made most scientists unwilling to become involved in the scientific effort aimed toward the next war.

The attitudes toward science and war of the political leaders of World War II were formed during the Great War. Winston Churchill, First Lord of the Admiralty during part of the Great War, had tried and failed to introduce the tank and other motorized devices to the battlefield. Never a student of science, he had a reputation for being intrigued by gadgets but seldom troubling himself to learn how they worked. Franklin D. Roosevelt, the American assistant secretary of the Navy during the Great War, believed he needed no scientific training to do his job properly and was only marginally interested in science. But he did back the efforts of a well-known physicist, W. R. Whitney of General Electric, to create a "bynaural audition" sound-detecting device to locate submerged submarines, and when Thomas Edison revealed to the *New York Times* that he had a "secret" plan to make the country "invincible" during the war, Roosevelt successfully urged the Navy to hire Edison as scientific consultant.

The future leader most directly and perhaps most deeply affected by science in the Great War was Austrian corporal Adolf Hitler. In October 1918 mustard gas rendered him blind and helpless for four days; Hitler was considered lucky to survive. Thereafter he bore psychological scars from the experience, and his gassing is also thought to have reinforced Hitler's belief in the near-magical power of mysterious, science-based weapons of conquest. This was not a reasoned belief, for he had a categorical disdain for matters of the intellect; Hitler presumed that such expressions of intelligence as science were inferior to, and subservient to, emotion and the will. Hitler's ire was frequently focused on what he believed were the attitudes of research scientists: elitism, liberal humanism, deliberate naiveté, and distance from practical matters; he and the Nazis preferred engineers to scientists, machinery to theory. In the "Myth of the 20th Century," Nazi philosopher Alfred Rosenberg wrote, "Whoever blindly condemns 'technology' nowadays, and piles up curse upon curse against it, forgets that it is based on an everlasting German drive which would disappear together with the downfall of technology. That would deliver us to the barbarians, and leave us in the condition to which the Mediterranean civilizations once declined."

Joseph Stalin considered himself a scientist—as had Marx and Lenin, who claimed to use the scientific method, with its objectively considered facts and proofs, as the basis for deducing economic and social imperatives.

In the interwar period most practicing scientists came to believe that wars were the enemy of science, while at the same time becoming remarkably oblivious to the events of the political world. They focused on the sustained and fundamental ferment within science's own house: physics was undergoing a revolution, based on the quantum theory of Max Planck and Albert Einstein, which generated new explanations about the nature and motion of subatomic particles, and these revelations were beginning to transform the study of chemistry and biology. At the same time scientific instrumentation was becoming more sophisticated, permitting closer and more revealing examination of matter through the use of X rays, electron diffraction, and electrically powered centrifuges.

As research at the frontiers of basic science became more complex and its new explanations more difficult for laypeople to comprehend, basic scientists seemed to feel no obligation to make the public understand what they were doing, or to devise practical uses for their science. Rutherford envisioned no practical applications devolving from his work on the atomic nucleus, and said that if there were any, he would not bother pursuing them. Most basic scientists looked down on applied scientists and engineers, whom they viewed as doing tasks reserved for lesser minds; perhaps this attitude also reflected the revulsion many scientists felt at the havoc wrought by the technologies of the Great War. Laboratory researchers now preferred to believe that science was liberating and humanizing, and they did not want such ethical attributes perverted in order to achieve the most practical of goals—military ones.

The implied divide between science and the military was a relatively recent phenomenon. In 1609, when Galileo invented a telescope for looking at the heavens, he touted its military potential, writing the Seigneury of Venice that with the instrument, "One is able to discover enemy sails and fleets . . . two hours or more before he discovers us, and by distinguishing the number and quality of his vessels judge of his force whether to set out to chase him, or to fight, or to run away." He gave the device and plans for making other telescopes to the Seigneury in exchange for a thousand ducats and a lifetime appointment as Professor of Military Science. Galileo's contemporary Francis Bacon was the first to articulate a distinction between basic and applied science, contending that basic research increased our understanding of nature while applied research increased our power over nature. The disdain of pure scientists for practical research and their refusal to involve themselves in military matters grew steadily from the seventeenth century onward. Exceptions such as the work of Sir James Dewar to devise cordite, an explosive powder—and of Rutherford himself, on various military devices during the Great War—only proved the rule.

In the 1920s and 1930s, convinced that science's greatest triumphs were achieved in peacetime—a notion unsupported by historical fact—scientists of high repute refused to "prostitute" their talents to make their countries into more effective belligerents. Their unwillingness to perform military-related work was exacerbated by the professional military services' reciprocal deep disdain for what science could contribute to the conduct of warfare. Commanders were already having to yield increasing fractions of their authority to experts in communications, targeting, detection, ammunition choice, and the operation of airplanes. They wanted to yield nothing more. According to physicist Solly Zuckerman, the mutual mistrust of the military and science in the 1920s and 1930s reflected fundamental differences. While science was always in an "unsteady state" in which "nothing is enduring [and] nothing is sacred," the military habitually remained very much in a steady state and considered a great many things to be sacred. Case in point: decades after the introduction of cars, trucks, motorcycles, and tanks, many armies were having difficulty phasing out horse battalions. Military historian B. H. Liddell Hart noted that well into the 1930s the way the military reached decisions on the larger questions of strategy, tactics, and organization was "lamentably unscientific. . . . There are no means for the comprehensive analysis of past experience and thus no synthesis of adequately established data to serve as a guide in framing policy." An air force officer, writing in a British military staff college journal, attributed the military's lack of enthusiasm for the work of scientists to the inability of each group to understand the other's world and its rules:

> The scientist deals primarily with inanimate objects and natural laws; the soldier with men, their handling and discipline. The scientist reaches his decisions by laborious experiment, by exploring every avenue, deliberately seeking the unusual and unexpected. Time is seldom of great importance to him. The soldier must make his decisions quickly, since the lives of his men and his whole force may be in jeopardy.

"Where it is the habit of the scientist to question, it is that of the soldier to obey," Zuckerman wrote. And he added that the split between science and the military was also a matter of class, since "the profession of arms was always a gentleman's calling" while "social prestige" had not generally been "enjoyed" by scientists. Peter Medawar, a physician and biologist who spent many years shuttling between scientific and military establishments, also identified as a class phenomenon basic science's denigration of applied research; in terms of pure science, he observed, there was no intrinsic need to place a higher value on basic than on applied research, and to do so was "a conscious and inexplicably self-righteous disengagement of the pressures of necessity and use."

Medawar and Zuckerman believed that such snobbism was mainly a British vice. It was not. Widespread, it found expression in the serious splits between the basic and applied sciences, and between science and the military, that existed in every advanced society in the interwar years, splits that would affect in important ways those societies' military readiness for the global conflict that lay ahead.

SECTION ONE

The Interwar Years

| CHAPTER ONE |

Journeys Toward Conflict

During world war ii many scientists involved in the war effort looked back and realized that they had completed a journey from an interwar-years stance of doing basic research and advocating humanistic concerns, to performing practical work for military aims as an expression of patriotic duty and because it was necessary to their survival.

In Russia that journey began earlier. As the Great War was ending for most other belligerents, in the new Union of Soviet Socialist Republics a conflict of equal severity continued. During it a substantial fraction of scientists and technologists emigrated to the West. Vladimir Ipatieff, one of the country's premier chemists in the czarist era, remained at home to mobilize the chemical industry for the new government. As with most notable scientists, he did not join the Communist Party, though he did make sacrifices, losing touch with a son who opposed the revolution and agreeing to turn his patents over to the state.

Czarist-era holdovers like Ipatieff and the 1904 Nobel laureate in physiology Ivan Pavlov became the principal members of the Soviet Academy of Sciences, and their students and successors continued to come from the old upper-class intelligentsia. And for a while young Russians of promise were permitted to study abroad, as they had in the past. Physicist Pyotr Kapitsa apprenticed to Ernest Rutherford at Cambridge, where he invented an apparatus that liquefied helium gas more efficiently, enhancing the study of low-temperature physics, and made theoretical contributions in magnetism. Kapitsa fell in love with things British, dressing in academic tweeds, wanting to live with his family in a thatched cottage in a quaint village.

To supplement Russian technological knowledge and to quickly upgrade Soviet industries, the USSR purchased Western technology and hired foreign technological experts—French for aluminum, Americans for copper, steel, and iron. Lenin also formed a military-research partnership with Weimar Germany, which set up military training, research, and weapons-production facilities at nine Soviet locations. Tanks were made at one site, poison gases at another; at a third, nearly all officers of the Luftwaffe were trained. By the late 1920s, as Ipatieff explained in his autobiography, the climate for science and technology in the USSR began to change:

> After the death of Lenin and Dzerzhinsky (two men who understood the value of knowledge and experience) a new and stupid type of individual came into ascendancy . . . a type of man who, out of jealousy and vanity, could not reconcile himself to the fact that he had to take advice from people who had formerly belonged to the upper classes. As a result, there began a mass drive on specialists; they were accused of malicious activity and ill will toward the government and were tortured, shot, or exiled.

Attacks on scientific specialists were cloaked in ideology derived from Lenin. In *Materialism and Empirio-Criticism,* Lenin promulgated six basic premises for the conduct of science: these limited science's reliance on mathematics to explain natural and social realities and insisted that since science made errors, its conclusions must be considered relative. The guiding principle of science was to be the advancement of the interests of the masses. As Alexander Vucinich writes in a history of the Soviet Academy of Sciences, Lenin's premises meant that scholars "were expected to be ready to wage a war on, and to reject or modify, all theoretical ideas advanced by modern science that were interpreted as incompatible with Soviet ideology as articulated by Stalinist philosophers." The incompatible ideas included Heisenberg's uncertainty principle.

Stalin's attacks on specialists had multiple objectives: to curb defections of Soviets posted abroad; to prevent "subversion" of the Soviet system by foreigners masquerading as technical assistants; to develop indigenous technologies, so that in any new European war the Soviet Union would not be technologically weak; to force Soviet scientists to apply their talents to meeting the country's economic needs; and to bring the Academy of Sciences under Communist administrative control. These goals were on display at a show trial in which a group of Soviet engineers—among them Ipatieff's chief assistant, and his best friend—were charged with conspiring with foreign intelligence services and former owners of Russian enterprises to prepare the way for foreign military aggression. All were found guilty and shot. Ipatieff and his wife barely managed to escape to the United States.

In 1929 the government sought to double the size of the Academy, proposing forty-two new applicants, many of dubious qualifications but a third of them Communist Party cardholders or sympathizers. The eighty-year-old Pavlov and other senior Academicians objected, but the Academy accepted thirty-nine of the forty-two, rejecting only three Marxist social scholars. Even so, the government ordered a new election, and the scientific workers' union, VARNITSO, assailed the Academy as a haven for leftover aristocrats and the ideologically incorrigible. Older scientists were purged from the Academy and forbidden to hold scientific jobs; some were exiled, never to return; the teachings of prominent physicists and mathematicians were "placed in the category of idealists" to be scorned, according to Zhores Medvedev, while "the tone of many scientific . . . journals became sharp, clamorous, and sometimes plainly vulgar." Real science receded, pseudoscience advanced—particularly in genetics, some Soviet geneticists beginning to insist, as Marx had, that acquired characteristics could be inherited. Scientists who believed this was false were exiled, including the founder of Russian population genetics, S. S. Chetverikov.

To replace him, party-liners tried to lure back from Berlin his best-known student, Nikolai Timoféeff-Ressovsky. A man who picked up languages easily—he could address audiences in Russian, German, or English—Timoféeff was also able to swap ideas with physicists and chemists. He was ready to return until a letter warned, "Of all the methods of suicide, you have chosen the most agonizing and difficult. . . . Book your ticket straight through to Siberia." Timoféeff remained in Berlin.

The most egregious instance of replacing legitimate science with illegitimate came with the rise of a charlatan named Trofim D. Lysenko. An unschooled agronomist who liked to walk barefoot in the fields, he first came to notice because of the publicity attendant upon a lucky break with the early planting of peas in Azerbaijan. Another fortuitous planting, of summer wheat seeds presoaked in water—he called this "vernalization"—which matured successfully after a mild winter when nonsoaked seeds did not, he touted as the salvation of the Soviet Union in the wake of massive losses of food crops. When fully developed, Lysenko's theory sent legions of farmers through their wheat fields armed with tweezers to remove particles responsible for pollination, thus forcing wheat to crossbreed. Lysenko's articles were a porridge of code words, misused scientific concepts, and peasant lore, but he smartly seasoned the stew with Marxist terminology and diatribes against "reactionaries" who dared question his work.

Party-backed Academicians took administrative power in the Academy of Sciences. Yearly requests for funding from researchers then had to be accompanied by detailed plans that explained how the research would fulfill the requirement of "planned utilitarianism" and lead to direct economic or social benefit.

Here was a point worth debating in every country, for legitimate arguments could be made in favor of planning in science under any sociopolitical regime. When only limited funds were available, it made social sense to choose projects having the best chance to advance the wider society's interests; and in times of crisis, it was entirely fitting for scientists to agree to be governed by the same rules of relevance that applied to everyone else in a society. These very points were brought out at a seminal conference in Great Britain in 1931, where Soviet scientists stressed the inherent logic of planning. The notion was reinforced during a reciprocal visit to the USSR by British scientists J. D. Bernal, J. B. S. Haldane, Julian Huxley, and John Cockcroft, who were impressed by the progress made by Communism in the Soviet Union at a time when the ills of capitalism had mired the rest of the world in sustained economic depression.

The most cogent refutation of the need to plan research was made by Pavlov. He said that to write out in advance what he would do and what practical good might come of it would limit the avenues he could pursue as a result of making discoveries, so he refused to make a plan. Pavlov was too prominent for the Soviet government to gainsay, but lesser scientific lights were forced to devise research plans that made economic sense and passed ideological muster, and to stick to them no matter what the results.

Like Ipatieff, Terada Torohiko in Japan was a holdover from another era, a product of the samurai class. In 1915, because the mails had been so slow, his articles about his research on X-ray crystallography reached the offices of *Nature* months after similar ones by the Braggs, and so Terada missed sharing their Nobel Prize. Frustrated with Western science, he turned to investigating the acoustics of the bamboo flute, the rumblings made by the sea just before a storm, and only slowly worked his way to practical matters: sea waves led him to wave propagation, to oceanography, to meteorology, to geysers, to volcanoes, and, after the earthquake that flattened Tokyo, to seismology. Suzuki Umetaro, known as the first major Japanese scientist from the farmer class, fared little better interacting with the West. He discovered a treatment for beriberi in a derivative of rice bran, and although he was one of the earliest scientists to insist that trace elements were essential to human nutrition, Suzuki also lost a recognition race, to a Polish chemist who named the rice-bran derivative vitamin B-1.

If the West ignored Japanese scientists, Japan itself was oblivious to their importance as resources. Ishiwara Jun, an accomplished theoretical physicist who published papers dealing with relativity and quantum theory—in German—had his scientific career in Japan derailed because of an extramarital liaison; when it was made public, he was forced to resign his university chair. Nishina Yoshio, an electrical engineer who became a physicist and the colleague of Rutherford, Heisenberg, Dirac, and Bohr, on his return to Japan was unable to secure an appointment as a physicist at a university because his degree was in engineering, and he had to settle for a place at an independent institute.

Bucking this trend were Japan's resolutely practical scientific institutes. The clinching argument for the establishment of the Physical and Chemical Research Institute was the contention that though the cost was equivalent to building a battleship, the warship would become obsolete in ten years while the research institute would improve over time and yield more important results for war and for peace. In the 1920s and 1930s such institutes and the military shared the development of steelmaking, underwater propulsion, and radio communications, employing scientists in an era when the prestigious universities continued to put barriers in the path of academic advancement for young basic researchers.

The ascension of Hirohito to the imperial throne in 1926 gave all science in Japan an expectation of the best possible patron, for in addition to being a butterfly collector, the emperor professed himself a believer in science. Unfortunately, what he understood by science mirrored the notions of his teacher Sugiura Shigetake, who had studied chemistry in England, publishing some minor studies in the *Journal of the Chemical Society* before veering away from evidence-based work to elucidate a doctrine he called "scientism." Relying on such opaque aphorisms as "Seeds not planted will not grow" and "If mature, something is hidden," Sugiura used the language and the laws of science to define human relationships and rationalize the authority of the emperor; in one book he wrote that

> the founder of our imperial family and the emperor's ancestors are the ones in this country who have accumulated physical energy over the longest period of time. . . . Because there is a continuous line of descent that extends down to today, from the point of view of the principle of the conservation of energy one must conclude that in Japan they certainly have the most energy/power. . . .

Hirohito incorporated into his own attitude toward his country's scientific and military prowess Sugiura's distorted notion of science as upholder and justifier of authoritarian excesses.

The three strains of science in Japan—basic research, the well-funded practical-applications focus, and "scientism"—came together most darkly in the son of a wealthy family who became a doctor in the Japanese Imperial Army. Bluster and a talent for sycophancy aided Ishii Shiro's rise through the military ranks, as did his reputation as a womanizer and carouser. While he was doing postgraduate work in

Kyoto, an epidemic of encephalitis killed many people; Ishii petitioned his superiors to let him go abroad for two years to study biological warfare in ports of call from Singapore to Ceylon, Greece, the European countries, the Soviet Union, the United States, and Canada. Upon his return, Ishii perfected a water-filtration mechanism tough enough to turn river water into something potable. Promoted and appointed a professor at the Tokyo Army Medical School, he lobbied to begin biological warfare work, because, as he wrote, "biological warfare must possess distinct possibilities [as a weapon], otherwise, it would not have been outlawed by the League of Nations."

During the first years of the Depression, which came earlier to Japan than it did elsewhere, industrial companies were consolidated into cartels and those cartels subsumed to the requirements of the military, in an attempt to free Japan from economic chaos—and to support the 1931 invasion of Manchuria. A provocation known as the Mukden Incident gave to the Japanese military the grounds for convincing the civilian government that Japan must respond.

Many Japanese scientists wanted little to do with militaristic activities and tried to protest the Manchurian invasion. Kotaro Honda, a physicist who had made a militarily significant contribution during the Great War by designing a better method for making steel, tried as president of Tohoku University in 1931 to turn it away from industrial and military research. Honda and the small basic-science community were more interested in a series of lectures on nuclear physics by Yoshio Nishina, lectures that young researchers would later credit with spurring them toward lifetimes of study. Conversely, for Japan's applied-research institutions, the Manchurian adventure presented an opportunity to pry more support from the government: they told the House of Peers that Germany and Great Britain each spent the equivalent of millions of yen per year on research, while Japan spent tens of thousands. Huge increases in applied-research budgets were soon granted.

Ishii petitioned the army to set up a facility for biological warfare in Manchuria—not to research ways of protecting people from diseases but, as he wrote to his superiors, to "experiment [and] develop new weapons." In midsummer of 1932, Ishii and a supporting battalion rushed into the small village of Beiyinhe, evicted all the inhabitants, and began construction on a facility for making biological weapons and for testing them on humans.

During the Great War, when the American Navy hired Edison, the inventor formed a board of scientist-consultants, mostly from industry. In reaction, academic scientists formed the National Research Council. The Edison and NRC groups began competing for government funds. Two scientists who worked for the NRC during the Great War emerged from this experience with contrasting views.

One was Vannevar Bush. Among the tasks given to the NRC by the Navy was to develop new antisubmarine-warfare methods and to replace inadequate detection equipment. Future Nobelist Robert Millikan oversaw that effort, including the independent work of an engineer, inventor, and assistant professor at the Massachusetts Institute of Technology, Vannevar Bush. The son of a Yankee clergyman, Bush had been sickly in his youth, and while in bed for a year recovering from rheumatic fever he learned to do things with his hands and solved intellectual puzzles. At Tufts College in 1909 he made his first invention, a mobile surveying device that mechanically traced out a map as its bicycle wheels turned, but he could find no buyers or backers to manufacture it. By his sophomore year he was substituting for his math professor, smoking and drinking with the most ardent of fraternity boys, and becoming known for his practical jokes and as a track star. He tutored to earn his tuition, completed his bachelor's and master's degrees simultaneously, and, in another year, his doctorate of engineering from MIT. He then returned to Tufts as an assistant professor, continued to invent, and managed a radio laboratory for an on-campus manufacturer.

During the Great War, Bush's idea was to create a magnetic field around a wooden-hulled ship and use it to detect the presence of submarines; when the Navy refused to fund the work directly, he obtained permission from Millikan to work on it privately, bankrolled by the offshoot of J. P. Morgan's company that owned the radio laboratory. "Since I was not in uniform and took no government money, I was a maverick," he proudly recalled. The Navy tested the device aboard steel ships, then rejected it, saying the requirement of using wooden ships made it useless. The British bought a hundred devices, and three made it into action. From this experience, Bush later wrote, "I learned quite a bit about how not to fight a war [and] the complete lack of the proper liaison between the military and the civilian in the development of weapons."

Back at MIT, Bush switched into management, wrote a textbook, and formed the Raytheon company to exploit his own and others' inventions. He was known for three A's—being arrogant, autocratic, and ambitious—but was idolized by the brighter students. In 1925 he and a graduate student built a mechanical calculator; a second Bush "network analyzer" was good enough to win a medal from the Franklin Institute. MIT was strengthening its pure-science departments to complement its notable engineering side; Bush encouraged interchange between the faculties. Thus in 1931, when he and a colleague built a third calculating machine that could solve differential equations—the basis of calculus—there were MIT colleagues in several disciplines eager to use it. A civil engineer wanted to forecast how a bridge would sway in a gusty wind; the Army wished to predict the movements of an explosive shell; and a physicist sought to calculate the scattering of electrons in atoms. The press celebrated the Bush device as a "thinking machine," and today it is recognized as one of the earliest computers.

James B. Conant, a promising young organic chemist, spent the war years with the Chemical Warfare Service, a group involved with the most dastardly technology of that conflict, a technology that seemed to many chemists a perversion of their science. After the war Conant was pleased that he did not have to go into industry to make a career but could return to Harvard. There he continued his research but soon discerned he was not going to make great discoveries and decided to become an academic manager. Within a decade after the war he became the leading candidate for the presidency of Harvard.

Conant's conviction that others would outstrip him in pure science was testament to the growing strength and depth of the American scientific community. Visiting Chicago, Werner Heisenberg was gratified that a dozen gifted graduate students attended his lectures; in Berlin, he remarked, he'd taught such a class to one student at a time. British universities granted science doctorates annually to two dozen men and women; American universities now granted more than a hundred, and their quality was considered equivalent. Germany had one outstanding mathematics institute; the United States was upgrading several to superior level. The Rockefeller and Carnegie Foundations made serious commitments to basic science; the amounts the Rockefeller spent abroad—for instance, to build new institutes of zoology and physical chemistry in Munich—were more than matched by what it spent at home.

In contrast to the practices of European research institutes, later émigré Hans Bethe would note, in America "the experimentalist constantly discusses his problems with the theorist, the nuclear physicist with the spectroscopist." This was a strength of American science, as was the country's relative lack of class distinctions, and its breadth, derived from the increasing population size and wealth of the post–Great War United States.

Recent graduates of American science programs did not disdain employment in industry, which welcomed them in droves. Patents and practical applications emerged in a regular stream from the industrial laboratories of DuPont, Monsanto, and the Bell Labs arm of American Telephone and Telegraph. At Bell Labs physicist Clint Davisson investigated the diffraction of electrons by crystals, work that would win him a Nobel.

Almost none of the benefit of the upsurge in American scientific power, however, flowed to the American military during the 1920s and 1930s, when the country's armed forces suffered a disproportionately greater decline in strength, budget, and political backing than did those of the other democracies. Contributing to the decay in the United States were headlined revelations about industrial profiteering during the Great War and Americans' widespread wish to isolate themselves from the European virus of military clashes. Moreover, the military did not respond well to scientists; when Millikan offered the services of the National Research Council to a conference of generals, the assistant chief of staff for supply implied that there was little need for scientists to dream up new weapons or to suggest ways to use scientific breakthroughs—those matters would be handled internally by the armed services. Even when an American plane manufacturer demonstrated a prototype of a newly designed small, lightweight monoplane fighter to the Army and Navy, it was turned down; the design was sold to Japan, where it became the basis for the Japanese Zero fighter.

In the early 1920s, the French, British, and American scientific communities deliberately isolated German science based on its connection to the discredited government that had begun the war, had continued it to the point of the slaughtering of millions, and must now pay a price for aggression. Germans received few invitations to international conferences, and important journals seldom accepted their papers. Shortly, however, a prewar ideal surfaced once again:

the internationalization of science, its ability to surmount national boundaries in the search for universal truths. Among those who could not deny the need to learn from the Germans were the physicist Frederick Lindemann and the chemist Henry Tizard, Englishmen who had been students of Walther Nernst in prewar Berlin. Both had become British test pilots, but even after the Great War retained their fondness and respect for German science.

Lindemann was the son of an Alsatian father and an American mother; born almost by accident in Germany, when his mother was on a vacation, he was British, but to many in Great Britain he seemed foreign. His inherited fortune and imperious affect contributed to that impression. In many respects a Renaissance man of science, he grasped what was current in various fields and could reel off ideas for them; Einstein called him "the last of the Florentines." At Oxford he became known as "Prof": a skilled low-temperature physicist, an expert on the mathematics of spin, a vegetarian, a tennis player, a pianist, a vehement anti-Communist who wore bowler hats and carried an umbrella. "I can understand and criticize anything, but I have not the creative power to do it myself," he once said. He would arrive at scientific meetings in a chauffeured Rolls-Royce, make deliberately offensive remarks to spark discussion or so he could fashion quotable retorts, and would then wonder why not every colleague held him in high regard. He despised most people, tolerated some, and cultivated only a few, like Churchill, with whom he developed a friendship based on love of language, sport, a fondness for wild schemes, and conservative political views.

Tizard lobbied for Lindemann's appointment to the chair of Experimental Philosophy at Oxford and the directorship of the Clarendon Laboratory, then a poor cousin to the Cavendish at Cambridge, led by Rutherford; under Lindemann's direction the Clarendon's reputation and facilities improved steadily. An intimate remembered a typical conversation between Lindemann and Tizard, which turned into a heated argument about nothing more consequential than the best way to pack oranges into a crate.

A wiry man of deliberately drab appearance, Tizard was less histrionic than Lindemann and had few personal quirks other than drinking champagne from a beer tankard. The son of a naval captain, he had done good graduate work, but Nernst had not considered him brilliant. Rejected by the Navy because of poor eyesight, he joined the Royal Flying Corps and worked on new methods of flying that took advantage of aerodynamics. He was about to test a new plane when German bombers were reported nearing London; he took off to meet them, but his guns jammed, so he spent his time in the air taking observations of the Germans' speed and maneuverability. After the war he moved from research to administration. Able to talk with scientists and civil servants alike, he steadily accumulated responsibility, becoming rector of a college and a member of important supervisory and review bodies.

Tizard's reviews were sought by the Air Force, not by Britain's vaunted Navy. In every major nation the Navy was the most hidebound of the military services, perhaps due to the habitual isolation of ships at sea from the dictates of land-based supervisors or to being the service most encumbered by tradition. In earlier days the British Royal Navy had resisted the introduction of steam vessels, then of submarines and airplanes as ship hunters. The Royal Navy continued to build big battleships when most strategic thinkers believed that airpower would soon negate their potency. The Admiralty did set up a Department of Scientific Research and Experimentation to examine wireless radio, submarine detectors, mines, and torpedoes, but although this department had a facility adjoining the National Physical Laboratory, researchers at one felt they had little to say to those at the other.

Crucial innovations in design and functionality of military hardware remained difficult to capitalize upon while political leaders embraced peace and disarmament and while good scientists disparaged applied research and could do so with impunity—because nearly every year a Briton would cover his country with honor by winning a Nobel Prize in chemistry, physics, or biology.

Another problem in Great Britain were the class distinctions separating scientific researchers and laboratory workers. Exacerbated by the Depression, they were further highlighted by the birth of a British Association of Scientific Workers modeled on the VARNITSO group pressuring the Soviet Academy of Sciences. The BASW was also an instance of the increasing influence of Communist thought on the democracies, an influence made palatable because facts that might have undercut it were then unknown, such as the purges in the USSR and the deliberate starvation of 5 million peasants. Although men like Tizard and Lindemann were adamantly anti-Communist, other leading scientists such as Bernal and P. M. S. Blackett, who were Communists, and Cockcroft, who was not but who shared some of their views, continued to fuse their belief in science's ability to improve the world to a belief in Communism's power to do the same.

In France the nuclear physicist Frédéric Joliot disliked Communists and characterized his sentiments as socialist, as did his teachers, the physicist Paul Langevin and the chemist Jean Perrin, and Joliot's wife Irène, eldest daughter of Mme Marie Curie. Langevin had been a student of Pierre Curie, then made contributions to the physics of magnetism and anticipated in part Einstein's ideas on mass and energy; after Pierre's death in 1906, Langevin took over his position and was also rumored to have become Marie's lover. Langevin's son-in-law, Jacques Solomon, was another leading physicist. Perrin headed a physical chemistry laboratory adjacent to the Curies', and his son and Irène had been tutored together privately.

Starting in the late 1920s, Joliot—then at the Collège de France, near the Curie Institute—and his wife began to collaborate on experiments. Sister Eve Curie remembered them as contrasting only in their personalities:

> She was as calm and serene as he was impulsive. . . . She found it difficult to make friends, while he was able to make human contact with everyone. She took little interest in her appearance and dress, while he was good-looking, elegant, and always a great success with the opposite sex. In argument, Irène was incapable of the least deceit or artifice. . . . Frédéric, on the contrary, without yielding on anything basic, knew magnificently how to use his intuitive understanding to put his opponent in a condition to accept his arguments.

During the Great War, French scientists had assisted the military, but after the war the Ministry of War seldom consulted civilians. Langevin, who had worked on detecting submarines, was dropped by the military after publicly supporting the cause of French mutineers. Conservatives such as Admiral François Darlan did not like left-wingers such as Langevin and were afraid that civilians might leak military secrets. Because of concern for secrecy, a French scientific manager later wrote, officers would "extract the implied problems of pure science and give them over, in their crystalline, unworldly purity, to civilian scientists to be studied as such, and without any consideration of their application to military technique." When civilian scientists offered to study magnetic mines for the Navy, Darlan told them it was none of their business.

By the mid-1920s the strengths of German science had become so apparent that for other Western scientists to ignore it was detrimental to progress, since Germans were making multiple theoretical and experimental breakthroughs. The prowess of Einstein, Planck, and Heisenberg, along with a recognition of the Weimar Republic as a struggling democracy aspiring to join the League of Nations, reversed Germany's prior scientific isolation. The Rockefeller Foundation's largesse enabled the growth of remarkably productive German scientific institutes and the formation of strong links—exchanges of graduate students, visiting lecturers, international conferences—among German, British, French, and American scientific communities. A good example was the collaborative work on vitamins carried on by Hess in New York, Windaus in Göttingen, and Rosenheim in London.

Windaus won the Nobel for chemistry in 1928, a year after Heinrich Wieland won and a year before Carl Bosch, head of the I.G. Farben company—three successive Nobels for German chemistry. Waiting for their Nobels were Otto Hahn, a nuclear chemist, and Otto Diels and Kurt Alder, whose diene synthesis yielded dyes, drugs, insecticides, lubricating and drying oils, and plastics.

Perhaps the most potent scientific-commercial collaboration in Germany was that of Fritz Haber and Carl Bosch, whose joint process was the basis of the nitrates industry that brought great profits to I.G. Farben. Bosch was an engineer and an astute businessman in addition to being a good scientist, but he was also subject to episodes of depression that would render him unable to function. Like the scions of the Krupp and Siemens families, Bosch was active in the leadership of the Kaiser Wilhelm Gesellschaft (KWG), a senate of industrialists and scientists who directed research funds for physics, chemistry, biology, medicine, anthropology, agronomy, plant and animal breeding, and the like, to the many Kaiser Wilhelm Institutes that supplemented university-based research laboratories.

Though the German general public and political leaders believed that physics was esoteric and did not have the relevance to everyday life that chemistry or biology did, and even though by 1930 the number of positions for physicists was in decline, physics was still the crown jewel of German science. That jewel's best setting was Göttingen, whose institutes were led by men with towering reputations: James Franck, Max Born, and mathematician Richard Courant. Outlying physics institutes were led by the equally accomplished Planck and Peter Debye, and featured past and future Nobelists Wolfgang Pauli, Erwin Schrödinger, Max von Laue, and Gustav Hertz. An international conference on any facet of physics that did not feature a clutch of these men was not worth convening.

Above these physicists towered Einstein, winner of the 1921 Nobel and the epitome of German scientific excellence. Resident in Germany only infrequently, because he was in demand elsewhere as a lecturer, Einstein continued to be a friend and correspondent to a dozen German physicists and chemists of the first rank, a sounding board for their ideas, a champion of their work. A pacifist, he was known for suggesting that scientists withhold their expertise from the military as a way to forestall future wars.

Quantum physics was so revolutionary, of such complexity, and so rooted in advanced mathematics that it was impenetrable even to some older, accomplished physicists, particularly to two earlier German Nobelists, Phillip Lenard and Johannes Stark.

Lenard had received the second Nobel ever awarded, in 1905, and believed he was cheated out of another by an Englishman with whom he disputed priority in cathode rays. He developed a serious dislike of Einstein's pacifism, his lionization by the British, his theories, and his Jewishness. Lenard subsumed his scientific work to right-wing politics, becoming notorious for writing that "in reality scholarship—like everything else brought forth by men—is conditioned by race and blood." Stark was a closer contemporary to Einstein, and for a while he kept pace with Einstein and used Einstein's hypothesis about light quanta in his own work. In 1919, Stark's research on splitting spectral lines in an electric field won him the Nobel. But he could not explain the "Stark effect" through classical physics, and he resented Bohr's and others' use of quantum mechanics to do so. He, too, began agitation on behalf of right-wing political aims and to advocate an "Aryan physics" that rejected the abstractions, theories, internationalism—and the purportedly Jewish origin—of the physics of Einstein, Planck, and Heisenberg, though the latter two were not Jews.

Biology, medicine, and physiology were another strong suit of German science. From the tracking of bees by Karl von Frisch to the imprinting of geese by Konrad Lorenz, the genetics of Alfred Kühn, the insect physiology of Gottfried Fraenkel, the embryology of Hans Spemann, the botany of Walter Schwarz, and the human physiology–pharmaceutical work of Ernest Chain, Germany's excellence in the field of biology was internationally recognized. At the KWI at Berlin-Dahlem were Otto Warburg and Otto Meyerhoff, both Nobel laureates; each made breakthroughs in many areas of biology and trained phalanxes of future Nobelists. "The scientific achievements and discoveries of the Warburg and Meyerhoff laboratories were staggering," Kirstie Macrakis concludes in her history of German sciences. The quality of these men and the terrific facilities attracted foreigners like Nikolai Timoféeff-Ressovsky, who in Berlin embarked on what a historical article labels "a period of amazing scientific productivity" in which he made experimental and theoretical contributions relating genes and mutations to the processes of evolution.

The biological data rapidly being produced by researchers around the world regarding genes, mutations, and inherited characteristics gave renewed impetus to eugenics, the improvement by breeding of plants and animals. As evidence accumulated that many debilitating diseases and physical conditions were indeed inherited, the definition of eugenics broadened to include the selective improvement of the human race.

Since today eugenics is associated in the popular mind with the depredations of the Nazis, it is important to recall that in the first third of the twentieth century several countries and half of all the American states adopted "eugenic sterilization" laws to prevent "imbeciles" and some physically or mentally deformed people from reproducing, in attempts to interrupt the passage of undesirable traits to future generations. On the recommendations of eugenicists, the United States' 1924 immigration act had racially based quotas; there was a eugenics section of the League of Nations and distinguished professoriats in many nations. Germany elevated eugenics to significant status in 1927 by establishing a separate KWI for it that sponsored the work deemed "absolutely necessary . . . since inadequate and dilettante efforts in this area [by Sweden, the United States, and Great Britain] have to be countered."

For German eugenicists, preventing the degeneration of the population included the removal of undesirable *racial* characteristics as well as those associated with disease or deformity. Alfred Ploetz argued that the quality of the Teutonic race was being undermined by misguided humanitarian attempts to protect weaker members, and sought to prevent the poor from having too many children. A special issue of the *National Socialist Monthly* was devoted to Ploetz's seventieth birthday. Fritz Lenz and Eugene Fischer collaborated on an influential textbook that dismissed the power of environment to shape human behavior because "the roots of most evil lie . . . in hereditary defects." Such reasoning appealed to Hitler; the Nazi Party adopted biological determinism, saying that their program to heighten the purity of the Aryan race by cleansing Germany of Jews and other undesirables was merely "applied biology." Lenz lauded *Mein Kampf* and Hitler as the "first politician of great influence who has recognized hygiene as the central task of politics and who wants to throw his full weight behind it."

At a time when the military establishments of the Great War victor countries were static or in decline, Germany's military research was in high gear, partly in reaction to restrictions placed on traditional armaments for that country by the Versailles Treaty. During the 1920s and 1930s, much of the military research was handled by technical universities and by industrial companies such as Siemens, Telefunken, Zeiss, and Daimler. Even within the ivory-tower communities there was plenty of applied, militarily relevant research: at Göttingen an aeronautics institute flowered under the direction of Ludwig Prandtl and his protégé Theodore von Kármán; it dealt directly with the Luftwaffe and with the commercial aviation manufacturers Heinkel and Junkers. Though the split between basic and applied researchers was as deep in Germany as in Great Britain, there were in Germany mediating factors, traditions of unquestioning obedience to authority and of submission to the primacy of the state, that found expression in the widely held belief among German scientists that when called upon by their government, they must willingly work in the service of their government's objectives.

The award of the 1930 Nobel in chemistry to Carl Bosch raised the status of the applied scientist in Germany above what it was in the rest of the world. I.G. Farben, the largest industrial employer of chemists, was known as a forward-looking and highly nationalistic company. However, as the Nazis drove toward full political power, they targeted I.G. Farben, calling it dominated by Jewish directors and officers. Bosch became greatly concerned as to how the firm might fare under a Nazi regime and dispatched an employee with good Nazi connections to convey to influential party members that Bosch and other directors were good Christians and that the industrial combine was vital to Germany's military future. Thereafter the Nazis softened their attack.

Nazi Party membership continued to grow along with the numbers of the German unemployed. By 1929, Hitler and his allies controlled 90 of the 491 seats in the Reichstag; in 1930 the Nazis had 107 seats and their allies an additional 41.

Bosch then sent another emissary to sound out Hitler about I.G. Farben's planned attempt to hydrogenate coal to produce oil and gasoline. Hitler spent several hours in the meeting and impressed the emissary with his willingness to learn technical details about the process. Hitler was enthusiastic about the potential of synthetic gasoline; he attributed Germany's defeat in the Great War to the blockade of raw materials and said he was determined that Germany become self-sufficient in oil, rubber, and other raw materials needed for truly mechanized armed forces.

In the early 1932 elections Hitler polled more than a third of the vote but lost to Field Marshal Hindenburg, though Nazi Party candidates swept many local contests. After the Nazis won 230 Reichstag seats in July, it became only a matter of time before Hindenburg would have to choose Hitler as the next chancellor.

Bosch arranged to meet Hitler in early 1933, shortly after Hindenburg anointed him as chancellor. The Nazi Party did not yet have an absolute majority in the Reichstag and was seeking a new election. Bosch quickly won assurance from Hitler that under Nazi leadership the German government would back the I.G. Farben synthetic-oil project and would look forward to similar projects with synthetic rubber. But Bosch then took the opportunity, as a ranking official of the Kaiser Wilhelm Society, to plead with Hitler not to force Jewish scientists to leave the country, which Bosch said would set back German physics and chemistry a hundred years. "Then we'll work a hundred years without physics and chemistry," Hitler shouted and had Bosch immediately shown the door.

On February 20, 1933, in a crucial meeting, Hitler addressed an audience of twenty-five leading industrialists; four of the chairs were filled with I.G. Farben directors, though Bosch was absent. Hitler told the assembled men that the election on March 5 would be Germany's last and that the Nazis would consolidate power either through being swept in at the polls or "by other means . . . with other weapons." The arms manufacturer Gustav Krupp understood that they were being told to bankroll Hitler's election or risk having their factories nationalized and businesses bankrupted. He pledged 1 million marks, and the other industrialists put up 2 million more.

In the March election Hitler did not win a majority of the Reichstag, but by preventing newly elected Communist delegates from being seated, he managed to push through legislation that enabled him to change the country from a republic to a dictatorship. By March 23, 1933, Hitler's power over Germany was absolute.

| CHAPTER TWO |

Germany, 1933–34

IN THE SPRING of 1933 Wernher von Braun turned twenty-one. A tall, handsome patrician, von Braun was studying for his engineering degree and was in love with the infant science of rocketry. Apprenticed to the German army division that under Walter Dornberger built rockets, in his spare time he took horseback-riding lessons with a Nazi-affiliated group and flying lessons from another such group but had not become a member of the Nazi Party. His father, Baron Magnus von Braun, recently resigned from his post as minister of agriculture in the last Weimar government, had retired to his farm in Silesia; the baron implored Wernher to get away from activities associated with the Nazis and to stay with the family in Silesia until Germany had rejected Hitler. Wernher did not yield to his father's entreaties, because the Nazi takeover assured that the military would play an ever-larger role in Germany, and the military's willingness to support rocketry was opening up the greatest opportunity of his young life.

Von Braun could not have joined the Nazi Party in the spring of 1933, because so many Germans wanted a party card that the Nazis declared their ranks closed; only "old comrades" would be able to benefit from their association with the new German leadership. The fewest old comrades were to be found on the faculties of the universities and Kaiser Wilhelm Institutes.

The Nazi Party had plenty of student members, but the professoriat had largely remained aloof from politics, neither embracing the party's rise nor actively resisting it. In that professoriat were thousands of scientists whose lives the Nazi assumption of power would adversely affect, just as surely as it positively affected von Braun's life. During their first months in power, the Nazis issued two edicts that directly touched scientists. The Law for the Restoration of the Career Civil Service, of April 7, 1933, commanded all institutions even faintly connected to the government—including post offices, research institutes, and universities—to rid themselves of "non-Aryans" and others considered undesirable or potentially disruptive because of behavior or background. It made exceptions for veterans who had served with distinction in the Great War or who had obtained their positions prior to 1914.

The Law on the Prevention of Genetically Diseased Offspring permitted the sterilization of people known to have serious genetic flaws, such as birth defects, idiocy, and insanity. Under this directive 350,000 people were eventually sterilized against their will and a path was set that very quickly led from sterilization to euthanasia, from the identification of people with genetically borne diseases to the identification of racial traits that the Nazis deemed defective, and then to killing of Jews, Gypsies, and members of other "races." This directive would ultimately enmesh many of Germany's physicians, medical researchers, and biologists in undermining and perverting the traditional helping role of medical personnel in a society. Eventually half of Germany's medical doctors would join the Nazi Party, a percentage of professionals far greater than that from any other discipline. Most of these physicians adopted and fostered the belief that Hitler was not only the leader of the nation but also chief doctor of the German *Volk*, the entity they imagined to be their primary patient. Such doctors' scientific integrity was wholly compromised; they believed in healing through killing and that they must destroy carriers of "diseased" genetic material as a "holy obligation to the state."

The Civil Service edict had the most immediate and direct impact on Germany's scientific research community. Hitler had found it intolerable that although Jews made up only 1 percent of the German population, they held one-eighth of all university chairs and had won one-quarter of Germany's Nobel Prizes. The edict forced the dismissal or resignation of 1,200 scholars during the first two years of Nazi rule, most of them in the first year.

For centuries German universities had been notably intolerant of nonconformity, and removal of professors for political reasons was a regular feature. Also, the Nazis' 1933 purge was not only directed at universities but was part of their larger attempt to rid Germany of Jews, Communists, and independent thinkers.

Max Planck believed that the edict was potentially devastating to German science. Known as the dean of the scientific community, the director of the Kaiser William Gesellschaft, and the winner of a Nobel Prize for his elucidation of quantum theory, he was not Jewish and had been a passive supporter of the Nazi "revolution"; never a believer in democracy, he had welcomed Hitler's ascension to power as a route to rescuing Germany from economic doldrums. But the edict so alarmed Planck that he obtained an interview with Hitler early in the new regime. In their meeting, Planck protested that removing the Jews would decimate German science and set it back for decades. Hitler replied, "Our national policies will not be revoked or modified, even for scientists. If the dismissal of Jewish scientists means the annihilation of contemporary German science, we shall do without science for a few years."

Of the 1,200 dismissals from the universities and independent research institutes, 400 were of physicians and about 100 each came from physics, chemistry, and biology. The remainder taught non-scientific subjects. A great many of those scientists forced to emigrate moved to the United States, Canada, and Great Britain, where most eagerly contributed to the scientific war effort. The removal of 15 to 20 percent of the established scientists in Germany enabled an entire new cadre to receive promotions: assistant chairmen took over departments, lecturers became full professors, assistant docents became lecturers. Many of the newly elevated people were seriously inferior scientists—"fools and charlatans," German historian of science Benno Müller-Hill has called them. Otto Kuhn, though an incompetent biologist, became director of the Cologne Institute of Zoology. Heinrich-Jacob Feuerborn, a Nazi Party member who lectured on heredity and racial biology, replaced a well-respected though half-Jewish man as the chair in zoology at Münster. Ernst Lehmann, a botanist at the University of Tübingen with only modest qualifications, because of his enthusiastic support of "the cleansing of the *Volk* from people of foreign races" became chairman of the German Association of Biologists and dean of the faculty at his university.

According to Ute Deichmann, author of *Biologists Under Hitler,* not a single German biologist "declined to accept a position that had been opened up by the expulsion of a colleague." Even those new appointees who were not Nazis or passive supporters of the regime could not avoid realizing that they owed their upgrades to politics, which thereafter made them less likely to challenge the Nazis and more vulnerable to what Müller-Hill calls "strong political and social pressure to toe the line politically [and] to a growing moral unscrupulousness."

Only a handful of principled, non-Jewish professors in various disciplines resisted the directives. There were also a few scientists, born Jewish, who refused to take advantage of exemptions that would have permitted them to maintain their positions. The most notable was Fritz Haber, the man who had fixed nitrogen and concocted poison gases in the Great War. Long a convert to Christianity, Haber was nearing retirement age in 1933, but in reaction to the Civil Service decree he resigned his post as director of the chemistry section of the Kaiser Wilhelm Institute. In a farewell letter Haber proudly claimed that for twenty-two years he and the KWI had "brought things of value both to science and the defense of our land." He left for Switzerland.

Albert Einstein had been under attack from the Nazis since the 1920s; he was out of the country when the Civil Service law was passed, and he concluded that a return would be dangerous; he considered offers from a half dozen countries before settling at Princeton.

Carl Bosch, the Nobel winner and the chairman of I.G. Farben, was not Jewish, and he made no public protest at the loss of Einstein, but he was upset by the ousting of Fritz Haber. Bosch attempted to use his position to have the Haber decision reversed; he also tried to form a committee of non-Jewish German Nobelists to lobby for Haber's reinstatement. Neither action had any effect. A dozen middle-rank chemists thrown out of their university or KWI positions received offers from I.G. Farben to enable them to remain in Germany; only a handful accepted.

Bosch's loyalties to Germany—and to I.G. Farben—overrode his distaste for the Nazis. He continued to entreat the Nazi leaders to support I.G. Farben's synthetic-oil project. They did, and they also involved Bosch's company in a secret plan to make aviation fuel and engine lubricants from coal for an Air Force whose existence was a direct violation of the Versailles agreement. By the fall of 1933, Bosch had successfully petitioned the government to resuscitate another languishing I.G. Farben science-based process, the manufacture of synthetic rubber, called buna. In these several arrangements the government guaranteed I.G. Farben a profit on everything it sold, even the products left over after filling the government's quota, which the company would sell to commercial users. Hitler personally brushed aside objections from Hjalmar Schacht, the chief economics minister, who worried about the budget-busting aspects of the agreements.

From exile, Fritz Haber attempted to prevent the transformation of the Berlin KWI for Chemistry that he had headed into an institute that would serve the military aims of the Reich. Since the Great War he had refused to do any research that had direct military applications. Now he tried to use as leverage the equipment that the KWI had purchased with the aid of Rockefeller funds, saying that the donor would not want it used for "the study of chemical warfare." Planck agreed and ordered that Haber's lab equipment be sent to England. The government vetoed this notion and appointed to the former Haber post a practical chemist whom Haber disparaged as a "perfectly unknown man from Göttingen." Over the objections of Planck, the Nazis also insisted that Haber's institute be redesigned for "military-technical purposes that lie in the framework of the Versailles treaty," an area, they contended, in which the new man had "already earned special merit."

A few months later Haber died in exile. His death occasioned one of the few public protests by the German scientific community against the depredations of the Civil Service Law. Planck organized a memorial service—an event deeply opposed by the authorities, with sanctions threatened for those who attended—at which he, Bosch, physicists Max von Laue and Max Delbrück, and other non-Jewish luminaries spoke of the loss to the scientific community of their Jewish colleagues. Von Laue's speech ended with a quote from Galileo, "And still it moves," implying that as Galileo's truth had survived the Inquisition, so now would the truths of Haber, Einstein, and other scientists survive exile by the Nazis. Later von Laue would block the nomination of "Aryan physics" founder Stark to the Prussian Academy of Sciences, an equally ineffectual act of defiance.

Physicist James Franck, winner of the Nobel in 1925, had offers from abroad but didn't want them. Born Jewish, he also didn't want the exemption to which winning two Iron Crosses entitled him. His resignation from Göttingen cited as grounds his belief that as chief of an institution he would be required by the new law to fire colleagues, which he would not do. Franck's resignation brought a storm of protest *against* him by forty-two members of the faculty, on the grounds that by resigning Franck was giving the foreign press ammunition for anti-German propaganda. Franck considered going into private industry but developed concerns about his extended family's survival. He went to Copenhagen to work with Niels Bohr, hoping to return when the Nazis faded away, then emigrated to the United States with his wife, daughters, and sons-in-law. Mathematician Richard Courant, the head of another of the Göttingen Institutes, who had also served in the war, tried at first to stay in Germany; his associates and even professional rivals signed petitions for his retention, to no avail, and Courant left for Cambridge and later for New York University.

Max Born wrote to Einstein from Oxford that it was easy for senior men like himself, Franck, or Courant to find a post overseas but added, "My heart aches when I think of the young ones." The "young ones," such as Born's former associates Edward Teller and Lothar Nordheim, started on the circuitous paths that would eventually lead them to the United States and to the Manhattan Project; similar destinations drew Courant's associates Hans Lewy, Herbert Busemann, and Werner Fenchel.

In the spring of 1933, when the Jewish-born Felix Bloch spoke about his fears of dismissal to his mentor and protector in the German physics community, Werner Heisenberg, Bloch later recalled, Heisenberg thought those fears were "somewhat exaggerated [but not] groundless." The pressure mounted, so Bloch quit before he could be fired. German-American mathematician John von Neuman of Princeton, who happened to be in Germany that spring, went around the country noting the names of physicists who would not be able to remain much longer; Bloch's name was on the list, which von Neuman circulated widely in the U.S. and in Great Britain. Bloch was offered a position at Stanford, and was embarrassed because he didn't know precisely where it was. Heisenberg said Stanford was somewhere near Berkeley, where J. Robert Oppenheimer was teaching, but that all he recalled about the California universities was that "they steal each other's axe." Later, Bloch would figure out that Heisenberg was referring to the Berkeley-Stanford football rivalry, in which each school would try to steal the other's symbol before the annual game. Bloch left for Stanford and never looked back, because, as he later said, "It was made so absolutely clear, particularly to the Jews who left Germany, that they were not wanted there anymore [and] could not very well have any serious thoughts about coming back." Nonetheless, they, and he, "deeply regretted" being "torn" from their German "roots."

Nobel laureates Gustav Hertz and Otto Warburg both remained in Germany. Hertz, with whom Franck had shared the 1925 prize, lost his post because of his Jewish descent and became director of a Siemens laboratory. The chemist Warburg, half Jewish, also remained in Germany through the entire Nazi period, in part because Hitler was afraid of getting cancer and believed that Warburg was on the verge of a cure.

Since highly talented scientists of Jewish heritage tended to gather in clusters, the Civil Service Law disproportionately affected university departments and research institutes in Berlin, where three-quarters of Germany's Jews lived, as well as a few other locations. Munich, near Hitler's base in Bavaria, had few Jews, but it had radical professors and those who expressed affinity with Communist causes, and so sustained quite a bit of damage. Among those fired was the mathematician and biologist Hugo Dingler, whose 1932 mathematical synthesis of genes provided a theoretical basis for the understanding of the double helix of DNA twenty years later; deprived of his positions, Dingler drifted from place to place in Nazi Germany, allowed to exist but not to teach nor to further develop or publish his work.

The impact of the edict was not felt with equal severity in all disciplines. Inorganic chemists were less affected than organic chemists, and chemists less than physicists, since there were more Jewish-born physicists. Many molecular biologists were forced to resign, but geneticists found their budgets rising.

Lise Meitner was preparing to teach a fall seminar in Berlin with Leo Szilard and Kallmann; the university severed all three from the faculty and canceled the seminar. Meitner's senior scientific partner of many years, the radiochemist Otto Hahn, resigned from the university but maintained his position at the Kaiser Wilhelm Institute. Max Delbrück, a non-Jewish physicist who had been an assistant to Meitner and who was working with the biologist Timoféeff, wanted to become eligible to teach classes but was required to attend an indoctrination seminar at an estate near Kiel. After three weeks, dubbed insufficiently sensitive to political necessities, he was refused a license; later he flunked another such seminar and knew he would never teach at a German university. Delbrück was grateful when a Rockefeller grant took him permanently to the United States.

Immediately after the elevation of Hitler, the Oxford Union Society passed a resolution that "under no circumstances will this house fight for King and country." An intimate of Hitler's later reported that Hitler greeted this announcement with satisfaction, since he construed it as the wish of the intelligentsia and expected the rest of Great Britain to follow. In 1933 while quite a few Britons had no inclination to meet whatever threat Hitler might present, British scientists responded with alarm and very quickly to the depredations of the Civil Service Law.

Like John von Neuman, Sir William Beveridge, director of the London School of Economics, happened to be in Austria and Germany in April, and witnessed firsthand the wave of dismissals in Germany. At home he organized the Academic Assistance Council, with Rutherford as its president and with Leo Szilard, a refugee who had recently been at the KWI in Berlin, as a key liaison to the displaced professors. Lindemann of the Clarendon Laboratory, who had trained in Germany before the Great War, undertook several trips to German universities to urge people he knew to leave, offering positions at the Clarendon to some. Their arrival accelerated the Clarendon's growth to the first rank of laboratories. Szilard also joined its staff.

The Royal Society provided offices and temporary bedrooms for many immigrants. An appeal to the general public, during an evening that featured Einstein, netted enough to award each professor and family £250. Later committees formed by England's Jewish community and by the Federation of University Women added maintenance grants, while the Academic Council worked to find the professors positions in universities, principally in Great Britain but eventually in thirty countries, including several in South America and at the nascent Jewish universities in Palestine.

In a half dozen other European countries, similar academic-assistance committees were formed, supported by local Jewish groups and by the Rockefeller Foundation, which, having provided important support to German universities and research institutions in the 1920s, responded with equal zeal to the task of finding jobs and funding for the displaced German professors throughout the world. The president of the Rockefeller Foundation, Max Mason, spoke German, and he and Warren Weaver, the head of the natural sciences program, had been in continual touch with German scientists before 1933; knowing how much help was needed, Weaver and

Mason committed the Rockefeller Foundation to give it. In France the moving force of the committee was the Russian-born Louis Rapkine, who worked at the Pasteur Institute; because he had not yet become a French citizen, and perhaps also because he was Jewish, Rapkine stayed in the background, and physicist Jean Perrin became the public leader of France's committee.

In the United States the Emergency Committee for Aid to Displaced German Scholars was led by Stephen Duggan, director of the Institute for International Education, with the assistance of the committee's executive director, Edward R. Murrow. The United States' severe Immigration Act of 1924 had an exemption for college-level teachers, but between 1930 and 1936 approximately as many American professors had lost their jobs as did German professors in 1933–34, though for different reasons, and the loss of American jobs for professors contributed to the hesitancy of American universities to absorb Germany's outcasts. Between 1933 and 1940, although the United States took in 100,000 refugees from German-speaking countries, it did not accept an additional 100,000, which its immigrations laws would have permitted. Among the fortunate to be absorbed by the United States was Emily Noether, considered the most brilliant female mathematician ever to have lived; she found a post at Bryn Mawr College.

In the United States the Duggan Committee and the Rockefeller Foundation took care to establish new professorships, rather than have the immigrants displace Americans from existing ones. They wanted to prevent a nationalistic backlash and similarly to avoid arousing anti-Semitic feelings that might emerge because almost all the newcomers were Jewish. In the United States, Great Britain, and France, anti-Semitism existed, and though it was mild when compared to the German form, it was pervasive and operated against native-born as well as against recent-immigrant Jewish scientists. Records show people from Jewish backgrounds being blocked at MIT, Harvard, Caltech, and other highly ranked American institutions. At most university science departments in the democratic nations, no more than one Jewish professor at a time was tolerated in each discipline. Yet the obvious need to help scientists from Germany raised the consciousness of many fellow researchers in the democracies to the evils of the Nazi regime and served as important impetus to those native-born researchers' willingness to cooperate with their governments by putting their scientific expertise in the service of Allied military objectives.

The second phase of the Nazi takeover of science was symbolized by the 1934 appointment of Bernhard Rust as Reich minister for science and education, a position for which he was scientifically unqualified; his background was political, his teaching and managerial experience minimal. Rust was notorious as the earliest official of a scientific organization, the Prussian Academy of Science, to have called for Einstein's resignation. Once elevated, Rust appointed as leader of the foremost funding institution, the Notgemeinschaft, Johannes Stark, the Nobel laureate and cofounder of Aryan physics, who had become an embarrassment to the legitimate physics community. Rust then hired Nazi Party officials and nonscientists as his deputies; the most egregious example of an unqualified man to take such a position was Rudolf Mentzel, who became head of the Reich Research Council, which administered the government's direct support of research.[1] These men forced a shift in the siting of the research away from the universities, which had formerly received 90 percent, and awarded more than 50 percent of the funding to the various Kaiser Wilhelm Institutes, many of whose new leaders owed their elevation to the Nazis.

They also began to separate German sciences into those that were useful to the regime and those that were not. Those of obvious prac-

[1]Mentzel was given Haber's substantial Berlin apartment.

tical or ideological use (biology, chemistry, geography, and engineering) received the most support; those that could not demonstrate their immediate utility (mathematics, physics, and psychology) were deprived of adequate funding, which made them more vulnerable to challenges by movements to Aryanize them—part of the push to make every science demonstrate allegiance to Nazi principles and practices.

In April 1934, just a year after the promulgation of the Civil Service decree, much of the transition had been accomplished. Max Planck was scheduled to make an address to open a meeting of scientific societies in Germany that was being held in an auditorium decked out with gigantic swastika flags. Physicist P. P. Ewald was in the audience and later recalled that "we were all staring at Planck, waiting to see what he would do," because

> at that time it was prescribed officially that you had to open such addresses with "Heil Hitler." Well, Planck stood on the rostrum and lifted his hand half high, and let it sink again. He did it a second time. Then finally the hand came up, and he said "Heil Hitler." . . . Looking back, it was the only thing you could do if you didn't want to jeopardize the whole Kaiser Wilhelm Gesellschaft.

| CHAPTER THREE |

Until the Shooting Began

WHEN GALILEO OFFERED his telescope to the Seigneury of Venice, both the inventor and the military recognized it as of immense military value. But just after the Great War, when Gugliemo Marconi announced that "it should be possible" to design a radio-wave system to permit a ship at sea to detect metallic objects like other ships and airplanes, Benito Mussolini, then in the process of taking over Italy, paid no attention. In 1922 researchers for the U.S. Navy discovered by accident that a ship passing between radio sending and receiving stations on the Potomac River would interrupt the signal and identify the ship's presence; but their proposal to use radio waves as a burglar alarm for a harbor was not funded. In 1925, Gregory Breit and Merle Tuve of the Carnegie Institution sent radio waves into the ionosphere (thirty miles above the earth) and measured how long it took them to bounce back, to determine the height and density of the upper atmosphere. Years later, Tuve would recall that during these experiments at Bolling Field, he and Breit would have to stop sending up pulses when airplanes were landing, because the planes interfered with the pulses. Had they investigated that interruption, they might have developed radar. Laboratories everywhere soon replicated the Breit-Tuve method of measuring the atmosphere but also did not make the leap to using radio waves to track the height of airplanes. In the 1920s, Kinjiro Okabe in Japan had discovered the efficacy of shorter wavelengths in distinguishing objects by radio waves, and noted physicist Yoshio Nishina had brought the military implications of a split-anode magnetron (a power source for generating radio waves) to the attention of the authorities. So had naval captain Yoji Ito, whose initial request to develop the magnetron was rejected by the Science Research Committee. Ito eventually convinced the Naval Technical Research Institute to fund it, but years went by before the Japanese Navy itself asked to have a radio-wave system, and they were interested mainly in preventing warships from colliding in the dark.

Radar would become the most important technology of World War II, and today we think obvious its abilities to spot obstacles in a ship's path or airplanes coming to the attack, and to guide gunfire. But it was not adopted as a military technology in the 1920s and early 1930s, for the most part because of the military's hidebound view that civilians—any civilians, and especially scientists—could not add much of value to the conduct of war. Another reason for the delay was the political climate, which in the 1920s assumed that there would never be another world war.

That assessment of the world's future began to change even before Hitler assumed power, and in 1933, when Marconi reiterated his suggestion about the military possibilities of using radio waves to locate distant objects, the Mussolini government responded by forming a special technical institute in Livorno to explore the technology and its military implications. The same changed political climate enabled France's Henri Gutton, who had been working privately to develop radar, to win permission to mount his machinery aboard the luxury liner *Normandie* on her maiden voyage. Though the set was judged inadequate, it did locate icebergs in the ship's path, and the data from this trial enabled Gutton to aim at generating shorter-length radio waves—shorter wavelengths, researchers were learning, meant better definition of the target.

The engineering branch of the Red Army of the USSR liked the idea of locating enemy aircraft by mechanical means, but they first tried using sound waves, which didn't work well, then infrared technology, which could pick out a plane only at night and in the absence of clouds. In the early 1930s, E. Oshchepkov, described as "a very able engineer" working with the civilian observer corps, received backing from the Academy of Sciences for radar research but ran into opposition from physicist A. F. Ioffe, who wanted to concentrate on generating longer waves, which then appeared best for directing ground-based guns, rather than on the shorter waves needed to detect faraway aircraft. Oshchepkov did convince Ioffe that their ideas were not mutually exclusive, and Ioffe's institute joined the work. Then Oshchepkov was arrested; his removal was a setback for Soviet radar, but it also resulted in research being transferred to the hands of the military.

In the United States, during the Depression, the Naval Research Laboratory, spurred by its director, Admiral Harold G. Bowen, worked on an early form of radar for ships; despite restrained budgets, the NRL work and Bowen's enthusiastic support of it convinced Congress to allocate $100,000 for the research.

In Germany in 1933 panoramic radar research was being done at the private laboratory of the Baron Manfred von Ardenne; he sought backing from Göring, but Hitler's second-in-command displayed little initial interest, perhaps because radar was a defensive system and the Nazis wanted only offensive ones. However, since the military itself had been told by Hitler to get ready for an eventual war, the

German Navy underwrote radar research by a Telefunken laboratory. Its pulsed radio-detection unit for ships was demonstrated for Admiral Erich Raeder in June 1935; impressed, Raeder asked for a prototype. Systems for gun-laying and long-range aerial detection were also being developed but were hampered because transmitting tubes had to be imported from Holland, and Hitler wanted Germans to rely on homemade materials. When Telefunken presented an early airborne fire-control radar to the Luftwaffe, the quartermaster general refused to buy it, saying it would "take all the pleasure out of flying." But even without the patronage of Hitler and his inner circle, by 1939 German radar in several forms was the best developed of any among the nations likely to contest the next war. Knickbein radar antennae at a handful of sites in northern Germany were ready to guide fighter planes to their targets; and the British, French, and Poles knew nothing about the Knickbein system, though the emplacements—30 meters high and mounted on turntables 100 meters in diameter—would have stood out on any aerial-reconnaissance photographs.

Radar was a technology that nonscientists had difficulty imagining. During a 1932 debate in the British Parliament, the lord president of the Council, Stanley Baldwin, stated that "the bomber would always get through." Great Britain's air exercises in the summer of 1934—in mock raids, both Houses of Parliament and the Air Ministry were mock-destroyed—seemed to validate Baldwin's contention. But in response to the 1934 exercises, Oxford physics professor and former test pilot Frederick Lindemann wrote a letter to the *London Times* saying that while there might be "at present no means of preventing hostile bombers from depositing their loads of explosives, incendiary materials, gases, or bacteria," that did not mean none could be developed; the problem was "far too important and too urgent to be left to the casual endeavours of individuals or departments. The whole weight and influence of the Government should be thrown into the scale to endeavour to find a solution."

Lindemann's own solution—which he did not describe in his letter—was a combination of untried technologies that did not include the use of radio waves. He imagined infrared sensors to warn of the bombers' approach and, to prevent the planes from loosing their bombs, massive antiaircraft barrages from below, combined with small mines containing high explosives being dropped by parachute into the paths of oncoming planes.

The physicist and his friend Winston Churchill (not then in the government but an influential writer and member of Parliament) pressed Baldwin to form a high-level air-defense committee. A. P. Rowe, a scientist advising the Air Ministry, had already cautioned that Great Britain was likely to lose the next war unless science found a solution to bomber attacks. Henry Wimperis used this alarum to spur the formation within the Air Ministry of a Committee for the Scientific Study of Air Defence, and proposed as its leader another former fighter pilot, Henry Tizard, and as members the antiaircraft expert A. V. Hill (a Nobelist in 1922 for work on muscular contractions) and the physicist P. M. S. Blackett (a former naval officer), along with himself and Rowe. Lindemann and Churchill were angered when they learned of that committee's chartering, ostensibly because it existed on what they considered too low a level—within the Air Ministry, which they distrusted, rather than at cabinet level—but also because Lindemann was not a member of the panel.

While political wrangling to get Lindemann on the Tizard Committee went on, in early 1935, Wimperis asked Robert Watson-Watt to respond to suggestions that a potential enemy might have a "death ray" of the sort depicted in science-fiction stories, to incinerate all living tissue in its path or detonate a bomb at long distance. Watson-Watt rather delicately phrased the question to his assistant in terms of the ability of radio waves to raise the temperature of eight gallons of water from 98 to 105 degrees, but both understood that it was an airman's blood that was of concern. Within ten days they reported to Wimperis that while a death ray was unlikely, using radio waves to locate an approaching enemy bomber was a real possibility. Thereafter, research proceeded with stunning rapidity. Within two weeks Watson-Watt's crew developed a working model and in another month were able to demonstrate its practicality to the Air Ministry at Orfordness—less than a month after Telefunken's demonstration for Admiral Raeder.

The Tizard Committee's demonstration carried more weight, because it directly addressed the fear of the bomber always getting through. Radar, then called RDF (radio direction finding), promised marvelous information on an attacker's position, height, speed, and path enough in advance to allow deployment of fighters before the attacker reached its target. By late summer of 1935 each passing month's tests were yielding new knowledge and better equipment. In September aircraft could be detected by RDF fifteen miles away, in August from twenty-five miles, and within six months, from seventy-five miles away.

For this small group of scientists, and for those who shortly joined them, developing radar was a wonderful way for them to help their country prepare for war: It was a defensive technology, not an offensive one, and it would save lives, not kill people more efficiently or in larger numbers. Even scientists who abhorred the idea of working with the military could understand and applaud radar.

Lindemann was then appointed to the Tizard Committee, and Churchill to an oversight Air Ministry committee. Despite the obvious success of the radar experiments, Lindemann disagreed with the direction of the Tizard Committee's work; he belittled the potential effectiveness of radar and insisted that the committee test parachuted aerial mines released by a drone plane.

Lindemann also insisted that more aircraft ought to be built before unproven defense systems. On this one Lindemann was right. As a sympathetic Tizard biographer, Ronald Clark, was forced to conclude, had Lindemann rather than Tizard been in charge of the committee, "it is likely that Bomber Command would have entered the war less ludicrously ill-equipped for the task at hand."

Through Churchill at the oversight committee, Lindemann mounted an attack on Tizard. A. V. Hill struck back in a poem that parodied Lindemann's rousing of Churchill to battle: "Von Alpha-plus" rides in his "shiny Rolls of eighty steeds" to the tent of "Odin, godlike son of destiny," to complain that Tizard's "professorial crew . . . will not be happy till the long-haired Greeks [Germans]/ Upon this city lay their infernal eggs." Von Alpha claims that the British were wasting "precious days . . . in vain experiment with R.D.F." But,

If, godlike son of destiny, we two
In place of Hopskipjump and Sigma were,
The sky would rain with parachuting mines
Unceasing, and the land be safe.

Lindemann's demand to be chairman provoked the resignations in protest of Hill, Blackett, and Tizard himself, which abruptly halted the activities of the committee. The Air Ministry accepted their resignations and established a new committee of Hill, Blackett, Tizard, and the two Air Ministry employees Rowe and Wimperis, but without Lindemann. This ended Lindemann's friendship with Tizard. According to a sympathetic biographer, Prof then "became a vagrant between the Admiralty, the Air Ministry and the Department of Scientific and Industrial Research," receiving "terse and almost contemptuous" replies to his letters full of ideas.

The Tizard Committee's new configuration still could not overcome the difficulty that Lindemann and Churchill had pointed up, of existing at too low a level and lacking political clout. Hill had to admit to Tizard in June 1936, that the committee "has failed as yet to achieve what it ought to have achieved owing to our inability to get large-scale trials carried out." Over the next three years, though, trials were made and a chain of radar stations established along the coasts, to give Great Britain warning in case of attack.

Three seminal events on the road to world war occurred in 1936: the Spanish Civil War, in which German and Italian planes, guns, and divisions assisted Francisco Franco's forces against Spain's government; Italy's aggression against Ethiopia; and the signing of the Axis pact linking Germany and Italy.

They stirred British Intelligence, after seventeen years of avoiding the task, to attempt to crack the new German military codes. During the Great War, Britain had broken the old codes and had thereafter assumed that publicity about that break would spur Germany to make new codes unbreakable, and so had not bothered trying to decipher them—another egregious failure of the imagination. By 1936 every intelligence service in the world knew that the German codes were being made by the Enigma machine, which had five sets of internal wheels and other devices that together could produce 159 million million million possible settings. That huge number seemed to nonscientists to guarantee the inviolability of the codes, but to a mathematician such a number was merely a challenge.

When the British asked their allies for decoding help, the French let them in on a secret. Marian Rejewski, a Polish mathematician, had been attempting to break the German Enigma and had made some progress because the Germans were changing the order of the internal wheels only once every three months, an error that gave Rejewski time to decipher some of the five-letter groups. The French had recruited a spy in the German Army cipher office and had been supplying him with money and female companionship in exchange for documents about Enigma settings and texts in both encrypted and plain form. With these, Rejewski was attempting to reconstruct an Enigma machine; he promised to supply a replica each to the British and the French intelligence services. One problem that he had already solved had to do with the keyboard. He had first assumed that since the Enigma had a QWERTYU keyboard (as did all typewriters) the letter Q would be connected with position #1 in the enciphering unit. After many failures based on that assumption, Rejewski figured out that the literal-minded Germans had not connected Q to position #1 but to position #17, because Q was the seventeenth letter in the alphabet. W was connected to #23, E to #5, and so on. Once Rejewski had made this imaginative jump, decryption proved easier.

The Spanish Civil War pitted forces from Germany against some from the USSR. Three years earlier Stalin had taken Hitler's elevation to power as precursor to a German invasion of the Soviet Union and began to prepare the USSR to meet it. His insistence on a sharp increase in funding for applied chemistry yielded new and improved fuels for internal-combustion engines, synthetic rubber, and insecticides and fungicides. He also strove to minimize Soviet dependence on foreign-technology imports and expertise, while doubling Soviet spending on science and technical research; by 1935 the Soviet Union was spending nearly twice as large a fraction of its gross national product on it as did the United States. "At a time when the capitalist world is closing down its scientific research . . . as a result of internal contradictions," a Soviet scientist gloated, research in the USSR had "completely unlimited" possibilities.

The touted Soviet commitment to science had brought to Moscow the German physicist Herbert Fröhlich and the American geneticist Hermann Muller. Muller had won international fame for studies on inducing mutations in fruit flies by X rays, for which he would later receive a Nobel, and had been working with Timoféeff-Ressovsky until Hitler came to power. A socialist and part Jewish, Muller avoided being thrown out of Germany by accepting an invitation to continue his research at the Soviet Academy. But in 1935 he and many other scientists in the USSR were brought up short by the official reception given to the plant geneticist T. D. Lysenko at a congress of collective farmers. In front of an audience that included government officials and Stalin, Lysenko defended his seed-softening (vernalization) program thusly:

> Was there not a class struggle on the vernalization front? In the collective farms there were kulaks and their abettors who kept whispering . . . into the peasants' ears: "Don't soak the seeds. It will ruin them.". . . . Such were the kulak and saboteur deceptions, when, instead of helping collective farmers, they did their destructive business. Both in the scientific world and out of it, a class enemy is a class enemy, whether he is a scientist or not.

Stalin responded, "Bravo, comrade Lysenko, bravo!"

After that there was no stopping the barefoot agronomist. He stormed upward through the scientific hierarchy while sabotaging the careers of legitimate geneticists and promising to develop high-yielding strains of crops in four years rather than the twelve projected by other scientists. The lionization of Lysenko made Herman Muller flee the USSR; he spent a year as a volunteer in the Spanish Civil War, doing work on blood transfusions, and wrote a book. Its theme: "Scientists have the responsibility of seeing to it that their efforts are used for the benefit, not disadvantage, of their fellow humans." Muller's thinking was an advance over the previous belief among scientists, that they were simply engaged in a search for truth and had no responsibility for the results of their research.

Lysenko and his acolytes, with the indulgence of Stalin, went after internal Soviet enemies, starting with Nikolai Bukharin, a member of the Academy and editor-in-chief of *Izvestia,* a former close associate of Stalin's who was now fighting the dictator. All were accused of being Trotskyites, German fascists, and racists. Kol'tsov, a geneticist who had made discoveries on the inheriting of blood diseases, was removed from the editorship of the journal he had founded and the institute he led, and was hounded to death. Tulaikov, the leading agronomist, was arrested and died in a gulag camp. Vavilov, the plant geneticist, was imprisoned and later died in custody; articles in Western scientific journals decrying Vavilov's treatment by the USSR were cut out of copies that the Soviet government permitted its scientists to read.

Only the intercession of Bukharin had permitted physicist George Gamow to attend a conference in Brussels in 1933. There Gamow pleaded with Bohr to find him a job in the United States; Bohr demurred and told Gamow to go back to the USSR. It developed that Bohr and Langevin had been instrumental in obtaining permission for Gamow to come to Brussels, and his refusal to return would undercut their influence; Marie Curie smoothed Langevin's feathers, and Gamow took the opportunity to emigrate to the United States.

He tried to warn Pyotr Kapitsa and his wife not to go to Russia without a written guarantee that they could come out again; Kapitsa thought that unnecessary because he was an international star, director of the Mond Laboratory, a low-temperature facility created for him at the urging of Rutherford. He had been making yearly visits to the land of his birth to renew family ties, but in 1934, when the Kapitsas tried to return to Cambridge, Pyotr was told he could not leave, as he had been declared essential to Soviet science. The authorities knew that Kapitsa was working on a device that used discharges of electricity to produce intense magnetic fields—which, as Kapitsa's patent application had put it, could be militarily useful in shooting a projectile from a gun with greater precision than could be accomplished by explosives. There was even speculation that Kapitsa's liquid helium work might be used to somehow make dirigibles impregnable, an idea that was scientific nonsense.

As with many fellow Russian-born scientists, Kapitsa was in sympathy with the general aims of the Communists and realized that the promotion of indigenous science and technology would require some sacrifices, but nonetheless he felt that his treatment at the hands of Stalin was counterproductive. Eventually Rutherford brokered a deal that permitted Kapitsa's wife to return briefly to England to fetch the couple's children and sent the equipment of the Mond Laboratory to the USSR, where it was reestablished in a new research institute that Kapitsa would lead. Shortly Kapitsa had a cottage built that was a virtual replica of an English village home.

In 1936 Kapitsa's colleague Lev Landau criticized A. F. Ioffe's Physico-Technical Institute for its low theoretical level and "primitive" experimental designs. Landau, a Bohr protégé who liked to spark disagreements, was arrested as a Nazi spy, though he was Jewish. Imprisoned without trial, Landau was freed only after Kapitsa threatened to resign. When Landau returned to the laboratory, he delineated theories about the behavior of helium in ultralow temperatures that decades later would him a Nobel.

Ioffe came under attack from VARNITSO for precisely the opposite of what Landau disliked: for spending too much time on theoretical physics and neglecting the needs of industry. Steering a middle course, Ioffe staved off the efforts of ideologues who wanted Soviet physics to reject the theories of Einstein, Bohr, and Heisenberg—and because of his tenacity, Soviet nuclear physics remained committed to theoretical purity and experimental objectivity, while Soviet biology suffered because it had no equivalent giant able to resist the trashing it received at the hands of the Lysenko group.

Vladimir Ipatieff, whose work in America was already transforming the petroleum industry and the use of plastics, now received a personal invitation from Stalin to come home and take up major responsibilities for the government. Then came cables from the son and daughter Ipatieff had left behind in the USSR, implying that they would suffer if he did not return. The threat was in some ways a testament to Stalin's view of Ipatieff's importance to Soviet militarization. Ipatieff refused to reenter the country without guarantees, and the Soviet Academy expelled him. That made him, he later recalled, "a scientific martyr overnight" and subjected the Soviet government to criticism for its "foolishness."

The inability of Soviet planes to outmaneuver German planes

during the Spanish Civil War unnerved the Soviet armed forces and pushed Stalin to insist that Soviet scientists apply their talents to such immediately practical tasks as helping design better planes. Those scientists who refused to act as researchers for industry and the armed forces were swept up in purges that decimated the professional military and the intelligentsia. There were so many executions and exilings in the last years of the 1930s that neither the Soviet military nor the scientific institutes recovered sufficiently by 1939 to adequately prepare the country for war. That insufficency helped tip the balance toward Stalin's decision to agree to the 1939 Non-Aggression Pact with Hitler.

After the Italians invaded Ethiopia in 1936, Vannevar Bush took note, and warned Americans that "the application of science to warfare will not cease in the world as it is now divided and governed." At a moment when outlawing war and universal disarmament were still being seriously considered, Bush was nearly a lone voice in the United States calling for the country to prepare itself for war and to utilize its scientific resources to achieve full preparedness.

Upon taking office, President Roosevelt had created a Scientific Advisory Board under Karl T. Compton, the Princeton physicist who had become president of MIT, but that board had not been able to do much. It identified such glaring problems as the Navy's inadequate salaries for civilian scientists, which had forced the Navy to allow civilian employees to patent their own inventions rather than deed them to the government, and which had practically halted the free cooperation between the Navy and commercial organizations like Bell Labs. The board recommended significant increases in government-supported military research: currently less than 2 percent of the services' annual budgets were being spent on R&D. Roosevelt brushed off that idea, believing that during the Depression any new money must be spent on social-relief measures.

During the mid-1930s Bush continued to invent and tinker with gadgets. He drew plans for a solar-powered irrigation pump but didn't bother sketching out an automatic telephone dialer. His most innovative work was on calculating machinery; in addition to improvements to the differential analyzer, he constructed a computing device that could sort reels of microfilm to quickly locate a desired piece of information. The "rapid selector" research led directly to devices to assist cryptographers in deciphering enemy codes. By 1936 Bush was pursuing a consulting contract with the Navy on that subject and attempting to forge better ties between that reluctant service and the engineers and scientists at MIT, where he had become second-in-command to Compton.

Bush and Compton had also became friends, buying neighboring farms in New Hampshire, sharing conservative political views and a vision of a high place for science and technology in the nation's affairs. The two notions were related: Bush and Compton agreed with columnist Walter Lippmann that the issues of the day were too complex to be understood by most citizens and should be decided by professionals able to judge matters in their respective domains "by right of superior specialized knowledge." Two other scientist/administrators shared these views: Frank Jewett, leader of Bell Labs, and James Conant, the organic chemist who at age forty had become president of Harvard. Conant liked Roosevelt's social ideas; the others did not. The four began to cohere through work on councils such as a Committee on Scientific Aids to Learning.

Today Pablo Picasso's indelible painting *Guernica* reminds us of the horrors produced during the Spanish Civil War by the strafing and bombing of German Heinkel 51s and Junkers 52s, which left 1,645 dead in a town of 10,000. In 1937, however, as much public revulsion was aroused by the use of poison gas by Japan in China and by Italy in Ethiopia. Many Western newspapers quoted an Abyssinian general's report on the dropping of mustard gas:

> Our people called it "the dew of death." Peasants would be working in their fields, women would be gossiping in the villages, chil-

> dren would be playing in the sun—all scores and scores of miles from the battlefront. . . . They would feel a moisture in the air. Then they would be coughing and choking, and their skins would be burning with ferocious heat that water couldn't cool, and they would be going blind and screaming.

Such news galvanized Sir Frederick Banting, a Canadian physician who had won a Nobel for the discovery of insulin. What frightened Banting was that chemical and biological agents could just as readily be released by airplanes. A dozen years earlier Churchill had described the threat of "pestilences methodically prepared and deliberately launched. . . . Blight to destroy crops, anthrax to slay horses and cattle, plague to poison not armies only but whole districts—such are the lines along which military science is remorselessly advancing." Banting wanted immediate research on defending against chemical and biological agents and on developing such agents, too, as a way to counter others' potential use. He had been an outsider to the scientific establishment, a physician who didn't obey the rules, and had almost been blocked from obtaining proper credit for discovering insulin. But when Banting raised the alarm about chemical and biological warfare (CBW) to the Canadian National Research Council, as John Bryden reports in his history of Canada's CBW program, the Canadian military immediately began "secret research" on both defensive and offensive aspects of chemical warfare, "apparently without consultation with either the government or Parliament."

During the Great War, American facilities that had produced massive quantities of poison gases, including a boatload of lewisite that was on its way to Europe when the Armistice was signed. The cargo was so deadly and dangerous that it was dumped overboard rather than returned to the United States. The remaining lewisite had been stored and almost forgotten in the intervening years, and there were plans to shut down all gas-making facilities in 1937 when the news from Ethiopia, and the urging of the Canadians, reversed that order.

In 1937, a year after signing the Axis pact with Italy, Hitler sought alliance with Japan on the grounds that they shared a potential enemy, the Soviet Union. An Anti-Comintern Pact that soon included Italy brought to Japan some scientifically advanced German technologies, including airplane design and code-making machinery. But the Japanese Imperial Army and Navy could not agree on a single standard for radar, and each service insisted on its own codes, which did not foster interservice communication.

Ishiwara Jun, the haiku-writing physicist who had been disgraced and lost his research position, became editor-in-chief of the leading scientific journal and wrote editorials decrying the "fascist" reorientation of science; Kotaro Honda, who once bolstered the Great War by improving steelmaking, refused after 1937 to have any further military-related research done at Tohoku University, where he was the president. In the wake of the 1936 military coup, most Japanese scientists scrambled for posts in which to hide until military fever had run its course. Previously, the mighty *zaibatsu*—the set of family-controlled industrial powerhouses in mining, chemicals, and manufacturing—had fought the Manchurian incursion because it brought international censure that halted the companies' ability to import raw materials. After the coup, though, the zaibatsu became alarmed at the government's awarding production contracts to newer concerns, such as those agreeing to make trucks for the military rather than automobiles for civilians, and started doing whatever the military wanted.

Good science went on despite the hostile environment. In 1934 Hideki Yukawa, a physicist who had studied only in Japan, postulated the existence of a subatomic particle that could explain the strong forces binding the nucleus, a particle later called the meson, for which he would subsequently win the Nobel Prize. With an eye toward providing cloud-chamber evidence of the existence of Yukawa's mesons, Yoshio Nishina completed the country's first cyclotron in time to coincide with a visit by Bohr in 1937.

That year a National Mobilization Law established a military-oriented Science Council to allocate funding for science and technology projects. "Basic cultivation of science" had been neglected for so long, the council's manifesto said, that at any moment "we might be attacked by a blockage [international embargo] of science and technology" that could prevent Japan from making "superior weapons" for proper defense. In response to the new law and council, most Japanese scientists kept their heads down and accepted war-related assignments but did not do them quickly.

When an outbreak of cholera killed 6,000 Japanese soldiers in Manchuria, Ishii Shiro was chosen to lead an operation to counter future biological attacks, to be disguised as a Water Purification Bureau, which the military also needed. Shortly, Ishii was giving demonstrations in which he urinated into a filter, then drank the treated product; once he offered his filtered urine-water to the emperor, who refused to quaff it. Ishii used the Water Purification Bureau as a cover to build the Ping Fan biological warfare complex, several thousand acres in Manchuria containing an airfield, a train line, laboratories, dormitories for workers and prisons for human subjects, a vegetable farm, 50,000 chickens to lay eggs in which to incubate typhus, and factories for producing both vaccines against diseases and toxins to spread diseases. Chinese prisoners were infected with cholera and plague, to test not the efficacy of vaccines but the course of infection and how long it took subjects to die. Ishii also tested poison gas, potassium cyanide, and electric shock. In hundreds of incidents in Manchuria, regular Japanese troops used poison gas; casualties were in the thousands. The gases were produced on an island named Okunoshima. In an attempt to keep the facility secret, that island was removed from official maps.

In September 1938, during the crisis in which Chamberlain gave in to Hitler's demand to annex the Sudeten districts of Czechoslovakia, the British government became so concerned about potential German poison-gas attacks that they issued 30 million gas masks to civilians. Adding to their worries was the threat of biological toxins; a recently emigrated biologist reported German trials of powdered anthrax and other toxins conducted with live animals on an island near Spain. Other intelligence reports told of the Italians' being able to produce 25 tons of mustard gas and 5 tons of lewisite per day, and of a similar massive effort by the USSR.

The British cabinet responded with an order to have British factories produce 300 tons of mustard gas each week. Aircraft were fitted with 250-pound spray tanks that could release mustard gas accurately from 15,000 feet. The French also began to manufacture an arsenal of 4.5 million grenades of mustard gas to be dropped from planes, as well as gas fillings for artillery shells and bombs.

In 1938 Hitler accelerated preparations for war, which he now said would come in three to four years rather than in a decade. His condensed timetable meant shelving designs for long-range bombers, since war would start before they could be properly developed; a crash program to make Germany independent of outside raw materials such as gasoline, rubber, and iron ore; and a changing of the guard in science administration.

Rudolph Mentzel replaced Johannes Stark as the head of the largest scientific funding organization; he restored research that the Aryan-physics founder had cut for ideological reasons, but he also appointed new divisional leaders such as Kurt Blome, a biologist who would later use his position for "research" in the concentration camps. Applications for funding were no longer extensively reviewed to determine their scientific worth. Basic researchers realized that the best way to obtain funds was to cite the danger, as one request from a non-Nazi put it, of being "outpaced by efforts abroad, in particular in the United States." Timoféeff-Ressovsky found it expedient to contend that an analysis of the diffusion of genetic characteristics would "support supervision in terms of racial hygiene." Such language from scientists of international repute helped embolden the Nazis' call for more drastic measures to rid the country of Jews and undesirables.

Timoféeff's research was led astray by this hysteria. Previously he and physicist Max Delbrück had been using physics concepts to investigate the nature of the gene; their experiments led Delbrück along the path that would eventually win him a Nobel in biology. Once Delbrück left for the United States, Pascual Jordan—a brilliant mathematician, a founder of quantum mechanics, an enthusiast of Freudian psychology, and a longtime Nazi—influenced Timoféeff to do genetics research based exclusively on the quantum mechanics concepts of target theory and energy movement. In letters Delbrück cautioned Timoféeff that this narrow, doctrinaire approach might lead nowhere, but he was ignored.

While some biological research in Germany remained unaffected by the rush toward war, everything that was faintly practical was reoriented to military goals. Erich Regener, director of the Physical Institute in Stuttgart, and his son, Victor, had been investigating the atmosphere by means of balloons, spectrographs, and Geiger counters; they were among the world leaders in this field. After Erich signed a 1936 Heisenberg memo denouncing Aryan physics and refused to divorce his Russian-born Jewish wife, he lost his post and funding. Victor escaped to the United States, but Erich appealed to Göring, who obtained for Regener his own KWI for the Study of the Stratosphere, which became the aviation ministry's research arm. It was still devoted to basic research, but the information gleaned about the upper atmosphere was also used to construct instrumentation permitting planes to fly at greater altitudes to avoid detection.

Carl Bosch found himself kicked upstairs at I.G. Farben, to a post having no operational responsibilities, as the company ousted its Jewish directors and made its Christian directors become members of the Nazi Party. Bosch also became the president of the Kaiser Wilhelm Gesellschaft. His episodes of depression grew more frequent as he found he could do little in either new position. I.G. Farben continued its transformation from a commercial company into a combine devoted to serving the military, making synthetic oil, aviation fuel and engine lubricants, buna rubber, poison gases, blood substitutes, magnesium, nickel, explosives, plasticizers, dyestuffs, and thousands of other products necessary to wage war.

The conversion of marvels to terrors was remarkably easy to accomplish. As a contemporary American textbook explained, "Many processes and equipment for making military and industrial products are practically identical, making it possible for certain plants to be changed almost overnight into munitions or war gas factories." Fertilizers, dyes, and engine fuels were the regular peacetime products of nitrogen, cellulose, chlorine, coal tar, and petroleum; but by combining these basic chemicals with readily available salt, sulfur, limestone, and grains, chemists could also use them to form explosives, propellants, poison gases, and nearly all the other compounds required by a war machine. Ethylene gas, made by cracking petroleum, could make either an alcohol or a mustard gas; by the addition of a few more readily available chemicals it could become ethylene glycol, which could be used either as an antifreeze in engines or as a replacement for glycerine in dynamite.

In the fall of 1938 a critical moment occurred in the history of nuclear fission. It demonstrated the truly international character of science and the importance of ideas unfettered by political or social claims. Back in 1919 Rutherford had shown that protons could be knocked loose from the nucleus of an atom, but through the 1920s the inner structure of the nucleus remained a mystery. In 1930 Walter Bothe and Heinz Becker determined that the nuclei of some lighter elements, bombarded with alpha rays, emitted radiation of stronger-than-expected penetration power, and Wolfgang Pauli predicted that the explanation of these results would involve a new particle in the nucleus. A flurry of experimentation began.

In France, Frédéric Joliot and Irène Curie aimed a stream of alpha rays at plates of carbon and metals in an ionization chamber, producing secondary radiation. Within weeks British physicist James Chadwick concluded that the Joliot-Curie secondary radiation consisted of fast-moving, uncharged particles torn from the nuclei, which he called "neutrons." Enrico Fermi in Rome and Ernest Lawrence at Berkeley also joined the chase. Joliot fretted over competition from Lawrence, who had a better cyclotron. But in mid-1933 the Joliot-Curies proved that the secondary radiation contained neutrons, protons, and positive electrons—the latter, a presumed product of transmutation.

On January 15, 1934, from a design made by both husband and wife, Joliot conducted the critical experiment, in which he bombarded radioactive elements with a new source of rays. At first he could not understand the results, so he called in his new assistant, Wolfgang Gentner, a German post-doctoral student who had become the laboratory's expert on keeping the Geiger counters in working order. "The counter is not all right," Gentner remembers Joliot saying to him. "There's something with the counter wrong." So Gentner, as he later recalled, "got a small gamma ray source and I tried out [the experiment] with gamma rays . . . and I found with these gamma rays that the counter is really working all right, and [the reason for the results] must be something else." Joliot finally figured out what had happened: He had transmuted one radioactive element into another one, producing artificial radioactivity. Joliot felt "a child's joy. I began to run and jump around in that vast basement." He and Gentner were so amazed that Irène was asked to come from her own lab to verify the results. Frédéric and Irène chuckled at the idea that at the Berkeley cyclotron "for a year already everything had been radioactive, but that had escaped them." The Joliot-Curies were correct, for when the Lawrence team read their report, one member wrote, "We felt like kicking each other's butts."

Enrico Fermi then published an article with the deliberately cautious title "Possible Production of Elements of Atomic Number Higher Than 92," which dealt with his bombardment of uranium by a stream of slow neutrons. An effort to produce "transuranic" elements began among the Fermi, Joliot-Curie, and Lawrence labs, and the Berlin team of Otto Hahn, Lise Meitner, and Fritz Strassmann. Fifty new elements were made within a year.

The Joliot-Curies won the Nobel Prize in chemistry for 1935. In his acceptance speech, Frédéric predicted that shortly, "Scientists will know how to bring about transmutations of an explosive character, like chemical chain reactions, one transmutation provoking many others. If such transmutations come to take place in matter, we can expect the release of enormous amounts of useful energy."

At a conference in 1937 Hahn told Joliot that while he respected Irène's work, he and Meitner had obtained data that contradicted Irène's transuranic results and would prove they were in error. When Hahn and Meitner couldn't do that, they redoubled their efforts to understand the phenomenon.

Lise Meitner had never denied her "non-Aryan" ancestry but between 1933 and 1938 thought the subject moot because she was Austrian and her family had long since converted to Protestantism. Experiments in the Hahn-Meitner lab had reached an acute stage in the fall of 1938, and she was reluctant to leave them, but during the Munich crisis Meitner finally understood that if she did not get out of Germany then, her life might be forfeit, and she let friends spirit her to Holland, then to Copenhagen to stay with Bohr, then to Stockholm. She was fortunate to have escaped. On November 9, 1938, Hitler's storm troopers wreaked the havoc of *Krystallnacht,* a pogrom against Jews throughout the expanded Reich.

On December 19 Hahn wrote Meitner of his own difficulties—he had been wrongfully included in an exhibition entitled "The Eternal Jew"—and went on to say that in the lab they had found "something so ODD . . . that for the moment we don't want to tell anyone but you." The odd thing: bombardment of uranium had produced what seemed to be barium, but half its weight, a result that, as Hahn soon put it in an article, "contradicts all previous experience of nuclear physics," and which he pressed Meitner to interpret. Meitner and her nephew, Otto Robert Frisch, who was working with Bohr in Copenhagen, went for a walk in the Swedish woods. Sitting on a snow-covered log, she sketched out her guess that the lighter barium might have been formed from the heavier uranium because the uranium nucleus had burst with tremendous energy. Recalling Einstein's formula—energy equals mass multiplied by the square of the speed of light—she calculated on a scrap of paper the energy that would have been produced by the bursting (or fission) of the uranium nucleus: 200 million electron volts. She was astonished by her own calculations. A Hahn-Strassmann article on the experimental findings was soon published, but the Meitner-Frisch article of explanation was delayed while Frisch tried to verify their thesis in his laboratory.

Frisch also conveyed his aunt's conclusions to Niels Bohr on the eve of Bohr's trip to the United States. Bohr became so excited that he had a blackboard installed in the ship so he could chalk out his calculations with fellow physicist Léon Rosenfeld. By the time they arrived in the United States, according to Rosenfeld, "Bohr had a full grasp of the new process and its main implications," which he quickly shared with the newly emigrated Fermi, and with Einstein and others. Bohr also became concerned that Meitner and Frisch would not receive proper credit for explaining the discovery. He was right, as credit went to Hahn and Strassmann; shortly Hahn, under increasing pressure from the Nazis who had taken over the KWG, would begin to claim that Meitner had had little to do with his discovery.

The idea of nuclear fission had become an express train that could not be halted. In Paris, Joliot quickly devised and completed what his associate Lew Kowarski later called a "brilliant and simple" experiment that proved fission was real and powerful. When Kowarski complimented him, Joliot said, "One shouldn't have any illusions. Just at this moment, the experimental proof of the same phenomenon is, no doubt, being obtained in other places." He was correct. Merle Tuve at the Carnegie Institution was then conducting an experiment in an apparatus that registered fission fragments as they were produced and was simultaneously telling a reporter on the telephone what he was witnessing: "Now, there's another one," Tuve said excitedly. Leo Szilard, the Hungarian-born physicist who had recently left Lindemann's Clarendon Laboratory to settle in the United States, conducted other experiments. On March 3, 1939, Szilard and a colleague began an experiment at Columbia University that could confirm the Meitner thesis. If they were able to corroborate the results, Szilard later wrote, "it would mean . . . the liberation of atomic energy was possible in our lifetime." The experiment was a success. "That night I knew that the world was headed for sorrow."

Szilard, Fermi, I. I. Rabi, and other physicists realized the military implications of nuclear fission and voluntarily agreed not to publish anything further about it. But Bohr published an article on "The Mechanism of Nuclear Fission," and though Szilard demanded that Joliot discuss these matters only "privately among the physicists of England, France and America," Joliot did not stop publishing either; his article offered proof that nuclear fission could be self-sustaining. Joliot's publication of that result in the spring of 1939 alerted physicists around the world and began an atomic-fission race. In their own attempt to win that race, Joliot and his associates wrote out secret patent applications for a crude uranium bomb and for a nuclear power reactor, and leased these to government in exchange for funds to continue trying to achieve a sustainable chain reaction before Germany could start a war.

Physicists Will Groth and Paul Harteck, after reading the Joliot, Bohr, Hahn, and Meitner articles, and though not fond of the Nazis, felt it their patriotic duty to inform the War Ministry of the possibility of making an atomic bomb. Similarly, in the Soviet Union, I. V. Kurchatov and others sent letters to the Academy of Sciences touting the possibility of making reactors and bombs.

Henry Tizard received a note about a possible atomic bomb, and there were proposals that the Air Ministry obtain a supply of uranium from the Belgian Congo and commission practical experiments. Tizard used a personal contact with the British director of the company that controlled the Congo uranium to try to place options on the ore being mined. The request was refused, but Tizard's alarm spurred the company to immediately send two shiploads of ore directly to the United States for safekeeping.

George Thompson, a Tizard colleague at the Imperial College, suggested they try to make the Germans think that Great Britain was already testing a uranium bomb, which might cause Hitler to hesitate before going to war. They readied a spurious memo with enough real data to impress, but the idea of letting the Germans see it was vetoed by higher authorities. Meanwhile, experiments on chain reactions continued under Thompson and also at Birmingham, where Mark Oliphant had invited Meitner's nephew, Frisch, to spend the summer. Oliphant and Frisch were soon joined by Rudolf Peierls. Their first results were not encouraging. Tizard argued in favor of continuing the experiments anyway but agreed with Churchill, who on the advice of Lindemann warned everyone not to take very seriously the possibility of making a bomb, because "there is no danger that this discovery . . . will lead to results capable of being put into operation on a large scale for several years."

On March 23, 1939, Wernher von Braun's twenty-seventh birthday, Adolf Hitler and entourage arrived at the Wehrmacht's rocket-testing center not far from Berlin. Von Braun, the technical director of development of long-range rockets, had joined the party in 1937. Walter Dornberger, von Braun, and Arthur Rudolph had been constructing and testing ever-larger rockets that they hoped would one day lift men into space but which they assumed would in the near future carry explosive warheads. At the time of Hitler's visit, no working rockets stood on the launch platforms, only engines and cutaway diagrams. With cotton stuffed in their ears, the visitors and rocketeers endured test firings of the motors that threw out blue flames and supersonic shock waves. At lunch Hitler recalled his early meeting with rocket pioneer Max Valier, whom he characterized as a dreamer, implying that Dornberger's nonworking rockets were no more realistic than Valier's. The rocketeers did not receive the backing for accelerated development that they had hoped to obtain.

In the spring of 1939 Vannevar Bush was fitting comfortably into his new post as president of the Carnegie Institution; he had taken the job in part because it was in Washington, D.C., enabling him to be more involved in the interaction between government and the private sector. He was also chairman of the NRC's Division of Engineering and Industrial Research and vice chairman of the National Advisory Committee for Aeronautics (NACA), positions that brought him into contact with the Army, the Navy, and the Army Air Corps. Although there were new airplanes on the military's drawing boards, Bush was concerned that not much else was being done to defend the United States from bomber attacks. No single agency coordinated work on antiaircraft guns, and as for radar, each service had pursued the technology separately and did not share information or resources.

Bush tried to remedy the situation by proposing a joint Army-Navy laboratory for antiaircraft research. The idea went nowhere. Wearing another of his hats, on behalf of NACA Bush lobbied Congress for a new aeronautics laboratory, arguing that while Germany had five laboratories, the United States had only one, and in consequence American planes had fallen behind Germany's in speed, range, and utility; the appropriations committee denied the request.

That spring two old friends met in Vienna. Hans Ferdinand Mayer was a forty-four-year-old physicist and engineer. He had been wounded in the Great War, took his doctorate in physics, became a research assistant to Lenard, taught at Cornell University, then returned to work for Siemens, in charge of what he later described as "one of the largest industrial research-and-development organizations in the communication and electronic field." His lab had switched from civilian to military work on "new and secret weapons which required communications and electronic techniques . . . There was hardly any secret weapon, conceived at that time [spring 1939] which was not known to me," he later recalled. That some weapons were already being manufactured and others field-tested had brought Mayer to the inescapable conclusion that war was imminent.

His friend was Cobden Turner, director of a British subsidiary of General Electric. Turner had become godfather to Mayer's second son, but the incident that had most cemented their friendship occurred in the fall of 1938, when Turner went to Berlin to rescue the daughter of Mayer's neighbor, who was in danger of being deported. Turner obtained a British passport and exit visa for Martyl, then took her home to become a member of his family. Turner's compassion deeply impressed Mayer, who decided in mid-1939 to tell Turner that if war were to begin, the peril to Great Britain from the new secret weapons would be immense. Turner asked what these weapons were. Mayer refused to give details but promised to provide information if war broke out. Mayer also gave Turner a small, very strong magnet, though he would not reveal its purpose. Turner's lab worked over the magnet and eventually determined that it was part of an explosive fuze, one that would burst an artillery shell, bomb, or mine only in close proximity to its target.

By 1939 the Germans were involved in many scientifically based secret-weapon projects, only some of which were known to Mayer. Forty that he did not know about were being tested at the Luftfahrt-forschunganstalt Hermann Göring, an aeronautics institute research facility in scattered buildings made to look like farm structures near the city of Braunschweig. The weapons being developed here were practical applications of the theory being taught at the nearby Braunschweig aeronautics and mathematics institute. Underground, in dummy buildings constructed beneath a thousand acres of trees, were specially constructed wind tunnels that could test planes and projectiles at speeds up to the speed of sound, with cameras capable of taking a thousand frames of photographs per second to study the workings of engines. The underground and concealed wind tunnels, as well as a rocket-testing facility, were constructed under the scientific supervision of Adolf Buseman, a former colleague of Prandtl and von Karman who was an expert in supersonic speeds and shock waves. The testing facility was purposefully sited in the countryside, Buseman later recalled, "because a lot of people who invented rockets died from the explosions, and therefore we couldn't build in the neighborhood of the town." In addition to testing planes and projectiles, the air force facility was test-firing a "vortex gun," designed to produce a wind vortex from dust that was exploded in the air with the expectation that it could bring down an airplane; a "wind gun" that mixed hydrogen and oxygen in explosive proportions to produce a sort of compressed-air and water-vapor plug that could act like a small artillery shell; and a "sound cannon" that used an explosion of methane and oxygen, focused by a parabolic reflector, to send a shock wave of sound that was supposed to be capable of killing a man in thirty seconds at a range of fifty yards, although it hadn't been tried on humans.

In May 1939 Reginald V. Jones found himself exiled to an outlying Admiralty Research station and vowed never again to have anything to do with Lindemann, Tizard, or Watson-Watt. Jones had studied under Lindemann for his doctorate in physics, then worked with the Tizard group trying to find new methods for defense against air attacks, then taken a post with RDF inventor Watson-Watt. Jones and Watson-Watt had quarreled, and the junior man was ostracized and exiled. It was a particularly trying time for Jones: He had recently become engaged and also had to care for his father, who was going blind. Thus when the secretary of the Tizard Committee telephoned to ask whether Jones would accept a new assignment, he had mixed feelings until he learned that the assignment was to find out what the Germans were doing "in applying science to air warfare." "A man in that position could lose the war—I'll take it," Jones said, and agreed to start combing the MI-6 files on September 1, 1939.

Before then he had a chance conversation with a former Oxford colleague, James Tuck, who was now working for Lindemann. Jones asked Tuck what new applications of science to war he should look for in his new job. "Reginald," Tuck said, "one day there is going to be a BIG BANG." Tuck expanded on this opening by informing Jones of the possibilities of nuclear fission and of the likelihood that the Germans were already working on it—conclusions he had drawn from an article by Siegfried Flügge, a nuclear physicist who seemed to be trying to warn the world that a nuclear bomb could be built in Hitler-controlled Germany.

The place to which Jones was to report, the MI-6 outpost, was soon relocated to Bletchley Park, two hours north of London. MI-6 had hired mathematician Peter Twinn from Brasenose College,

Oxford, and in the summer of 1939 added a more highly regarded mathematical theorist, Alan Turing from King's College, Cambridge, the man who had already developed the principle that would underlie the modern digital computer. The hiring of mathematicians initially bothered the classical humanists who had broken the Great War codes, but they were forced to concede that the near-infinite number of combinations made possible by the Enigma required the use of mathematics to decipher.

In late July 1939 Colonel Stewart Menzies, deputy director of MI-6, wearing a dinner jacket with the Légion d'Honneur in its buttonhole, went to Victoria Station to greet the arriving Captain Gustav Bertrand, head of the French code-breaking operation. Bertrand handed Menzies a replica of a German Enigma machine, made by the Polish mathematician Rejewski. This was the payoff for what Bertrand later described as an "incredible adventure, unequalled in any country in the world," to obtain the information for, and to actually construct, the Enigma. To Turing and the MI-6 cryptanalysts, the Rejewski machine was just what they needed; the only remaining uncertainty was how long it would take them to master the German Enigma, whose settings were now being changed very frequently.

On August 1, 1939, following successful demonstrations of airborne radar to the head of Bomber Command and to Winston Churchill, the British Air Ministry told E. G. Bowen's group to construct thirty radar sets and install them in Blenheim night-fighter planes before the end of the month. They would complement the Home Chain of ground-based radar stations along the coasts of the country, stations that were nearing completion.

In early August 1939 the Heinkel Company demonstrated for Erhard Milch of the Luftwaffe a test model of a jet plane whose airspeed far exceeded that of any propeller-driven plane. This was a masterpiece of technology, and the culmination of advances in the development of gas turbines and of steel alloys that could withstand temperatures up to 1,000°C and tension up to 30,000 revolutions per minute. The turbine engine produced the jet effect by compressing and then expelling heated air at high veolcity. The He 178 would take some time to develop to the point of mass manufacture, so Milch relegated it to the back burner while he and Göring dealt with the more immediate problem of preparing for a war that Hitler had scheduled to begin within days.

Werner Heisenberg was traveling in the summer of 1939. He had been cleared for the six-week trip by the SS, which cited his volunteering for military duty in 1938 as a reason that he could be counted on as a good ambassador of the Reich and to return home after venturing abroad. At the University of Chicago he stayed with Samuel Goudsmit, a Dutch-born physicist and the codiscoverer of an important principle of electron spin. The old friends discussed the exciting new research in nuclear physics, but Goudsmit did not sense that Heisenberg was intimately involved.

Shortly Goudsmit crossed to Great Britain for the annual conference of the "British Ass," the British Association for the Advancement of Science, where he met Walther Gerlach—it was Gerlach's experiment on molecular beams that Goudsmit and his collaborator had explained. Gerlach was evasive when Goudsmit asked what he was doing; Goudsmit concluded Gerlach must be working for the German government.

In New York, Enrico Fermi became nervous. Since the spring the Nobel winner had been trying to inform the American Navy of the possibilities inherent in being able to split the atom and the danger of Germany's doing it before the United States, but he had failed to make an impression. So during the summer he, Szilard, Einstein, and

Eugene Wigner decided to convey their concerns to President Roosevelt. They had also read the Flügge article and knew as well as anyone in the world the devastating potential of splitting the atom and that former colleagues like Heisenberg and Hahn were more than equal to the task of conceiving a bomb. They also imagined, more vividly than native-born Americans, the likely consequences should the atom's destructive power fall into Hitler's hands.

Einstein, ever the pacifist, refused to act as a direct emissary to the president, but he did write a letter on August 2, which was supplemented by a report of the work done by Fermi and by Szilard, and by a Szilard memo in layman's language on the possibilities of nuclear fission. Then they looked for an intermediary and for the right moment to present their warning to President Roosevelt.

The atomic physicists in the United States were still searching for the right way and time to approach President Roosevelt when, on August 23, 1939, the Third Reich and the USSR signed a Non-Aggression Pact. The compact of the dictators erased Hitler's fear that the USSR would join with Great Britain and France to oppose Germany's expansion and made it less risky for him to invade Poland on September 1, 1939.

SECTION TWO

European War

| CHAPTER FOUR |

War and Phony War

In response to Germany's invasion of Poland, on September 3, 1939, Great Britain and France honored their treaty obligations to Poland by declaring war on Germany, widening a Silesian war into a European war. Grand-Admiral Erich Raeder wrote in his diary, "Today there began a war . . . which—to judge from all the Führer's utterances hitherto—we should not have had to reckon with before about 1944; and a war which he considered until very recently he must avoid at all costs." For years Hitler had told Raeder that Germany would be able to take over Poland without having to fight the Allies.

Admiral Karl Dönitz, chief of Germany's submarines, understood that this wider war could not be won with conventional weapons. He wrote in his diary that "the only possibility of bringing England to her knees with the forces of our Navy lies in attacking her sea communications in the Atlantic. I therefore believe that the U-boat will always be the backbone of warfare against England."

The Reich's submarines, sent immediately to attack Germany's European opponents, made it abundantly clear this war would be highly influenced by science. The submarines carried a mix of very effective and quite ineffective science-based weaponry, a mix that owed as much to politics and personalities as it did to the state of scientific knowledge. To detect sonar the U-boats carried S-Gerät equipment—an advance beyond the previous generation of detector—but it did not work well (even though it had been chosen over a version of the surface Navy's Seetakt radar); their hydrophones were good but were easily damaged by depth charges, and their only defense against air attack was an ability to dive quickly. Dönitz railed to his diary about the "inadequate development and testing [and the] uncritical attitude by the Torpedo Experimental Establishment towards its own achievements." There were just 57 U-boats, a third in repair docks and a third not suitable for Atlantic waters; plans for 30 new U-boats a month had been plagued by production-line problems and managerial neglect. But even with not very good equipment, during the first weeks of war Dönitz's small force sank dozens of commercial vessels bound for British ports, a passenger liner, and a few Royal Navy ships.

In Poland, the German land and air offensive made only modest use of scientifically advanced weaponry: German tanks and other components of the mechanized assault ran on synthetic fuels and buna rubber; smoke grenades blinded Polish gun emplacements, and flamethrowers annihilated their defenders—neither weapon had been seen before—while dive-bombers frightened people with propellers that produced a screaming noise. But the blitzkrieg depended mainly for its success on swift movement, surprise, and overwhelming force: a million German soldiers either entered Poland or stood in reserve on the borders. By the time the Allies declared war, the military fate of Poland was all but decided.

At 11:00 A.M. on September 3, when the British and French ultimatums to Germany ran out and a state of war existed, London's air-raid sirens blasted. Winston Churchill's wife commented favorably on German promptness and precision as the Churchills hurried to a shelter. Churchill feared an immediate knockout bomber attack. The leading antiappeasement member of Chamberlain's party, he had finally been appointed to the War Cabinet as First Lord of the Admiralty; one of his first acts was to tap Lindemann to lead a statistical section for the Admiralty. German bombers did not materialize over London that day; a false alarm had been triggered by the appearance on a radar screen of a lone aircraft, that of the French air attaché returning from Paris; the nervous British reaction to the radar image included the scrambling of fighter planes, two of which collided in the air, killing all aboard.

For British scientists, the commencement of war broke the barrier that had previously kept them from working on weaponry. Those who had been reluctant to participate began to do so; others, who had worked on such marvels as defensive radar, shifted into work on offensive radar, for instance, the kind required to guide a bomber to its target.

On the second day of the war, the absence of that sort of guidance radar made a failure of the British conventional-weapons attempt to knock out the German fleet based at Wilhelmshaven. This raid epitomized Great Britain's own mixed bag of science-based war preparations. Lacking radar-directed navigational equipment, 5 of the 29 British bombers became lost and did not reach their target. The ones that did get to Wilhelmshaven dropped their bombs with an appalling lack of accuracy; the single direct hit on a German ship did not explode on contact, prevented from doing so because of an eleven-second delay fuze. Fourteen bombers were recalled out of fear that they would miss their targets and hit civilians—a reasonable worry, since an errant bomber killed two civilians in a Danish town more than a hundred miles from the intended target. While the first

British wave of bombers partially surprised the Germans, the second did not and ran into antiaircraft fire from the big German ships, which with the help of first-generation gun-laying radar caused considerable damage. Fuel tanks on the British bombers were so unshielded that a single bullet could bring down a plane, but the German fighters that rose to meet the Blenheims and Wellingtons were able to withstand repeated aerial gunfire hits.

In the aftermath of the Wilhelmshaven raid, British scientists previously unconnected to the military were now asked to examine planes that limped back from it and a downed German bomber. They discovered that testing done before the war by in-house staffs of military organizations had been wildly inadequate. While the Luftwaffe plane had self-sealing fuel tanks that enabled it to function even after repeated hits, the Royal Air Force plane's fuel tanks did not. RAF specifications had called for crash-proof rather than bulletproof tanks, and the testing procedure used to evaluate them had been remarkably unscientific: dropping a tank of liquid sixty feet from the roof of a building down onto concrete. None had survived that fall intact, and designs that might well have been bulletproof but were unable to withstand such a crash were rejected in favor of tanks that were relatively crash-proof but not bulletproof. New research led to the commissioning of tanks that contained a spongy rubber envelope inside, which could reseal after a bullet passed through.

Among the horrifying statistics that Lindemann had to present to Churchill was that in the first two weeks U-boats sank more than 20 commercial vessels and on September 17 the *Courageous*, whose 687 men were lost. Though a loss rate of 100,000 tons of shipping a month was not enough to cause serious economic damage to Great Britain, the steady depredations of the U-boats frightened Churchill.

The U-boats' success emboldened the German leadership. Admiral Raeder shelved his cherished ideal of a naval force balanced between surface and submarine vessels and demanded more U-boats.

Hitler agreed, but Göring, who believed that his Luftwaffe could win the war unaided, did not speed U-boat production. And so Allied fire and seagoing accidents continued to destroy as many U-boats per month as were being built.

The active phase of the Silesian war drew to a close after September 17, 1939, when the USSR invaded Poland from the East, advancing forty miles on a broad front without meeting opposition and effectively ending Polish resistance. On September 19, Hitler made a victory speech at Danzig.

According to a British Foreign Office translation of that speech, Hitler said that Germany had a secret weapon against which no defenses would prevail. A few days later Air Ministry intelligence liaison officer Reg Jones received his first urgent assignment: to look through the MI-6 files for clues to the secret weapon mentioned by Hitler. Both the Admiralty and the Foreign Office had candidates for the secret weapon. The Admiralty insisted it was an antimagnetic mine; the Foreign Office's candidate was a weapon to blind and deafen its victims. Jones resolved the puzzle by obtaining new translations: the Danzig speech had used the word *waffe*, which had been erroneously translated as "weapon" even though the context made it likely that the ultimate weapon to which Hitler referrred was the Luftwaffe. Another speech on which the FO had relied had also been mistranslated: Hitler did not want to blind or deafen his enemies but to render them "thunderstruck."

Jones made his report into a programmatic document by insisting that Intelligence files must thereafter be regularly consulted, to test the reality of the war against evidence in the files that Germany was developing bacterial warfare; poison gases; flame throwers; the combination of gliding bombs with aerial torpedoes and pilotless aircraft; long-range guns and rockets; new torpedoes, mines and submarines; and magnetic guns. "The war would see how many of these forecasts would come to pass," Jones later recalled. Every single one of these was developed by the Third Reich, and all but the chemical and biological weapons were used. Out of Jones's report came a proposal, which Tizard supported, that material gleaned from various sources about German secret-weapons development should be shared by all three armed services; this reasonable idea was vetoed by the Admiralty, which did not want to share its intelligence with the Army or Air Force.

The American Congress was on annual hiatus when the war had begun—Hitler had figured that recess into his plans, wanting to complete the battle phase before President Roosevelt had the opportunity of throwing the weight of the United States of America into the fray. On returning from vacation, Congress was in no mood to permit the United States to enter the war, and since by then the shooting had stopped, there seemed no overriding reason to do so.

Despite Congress's isolationism, Roosevelt began to assist the Allies and to increase the readiness of the United States for war. On October 11, 1939, the president was visited by an old friend, Alexander Sachs, who carried Einstein's letter and the Szilard and Fermi reports regarding splitting the atom and its ramifications for fashioning an atomic bomb. Perhaps the most effective document wielded by Sachs was a quote from another scientist who called the exploitation of atomic energy inevitable and suggested that if one country didn't do it, another would.

"Alex, what you are after is to see that the Nazis don't blow us up," Roosevelt said.

"Precisely," Sachs responded.

"This requires action," Roosevelt concluded. Ten days later the president convened an Advisory Committee on Uranium, chaired by the director of the Bureau of Standards, Lyman Briggs, and with Tuve, Szilard, Fermi, and Edward Teller among the members, to look into the subject with dispatch and under the greatest secrecy. But the research did not proceed very swiftly in the United States. Fermi and Szilard had recommended large-scale experiments to prove that a bomb could be made; however, in the fall of 1939 Fermi preferred to occupy his time with calculations about cosmic rays, and Szilard could not raise the funds or commandeer the facilities for the experiments.

Across in France, Joliot's group, which had the active help of the French government, obtained five tons of uranium ore and started work that was considerably in advance of similar attempts in Great Britain, the United States, Germany, and the USSR. Joliot accepted a commission as an army officer and was assigned to his own laboratory; his principal collaborators, Hans Halban and Lew Kowarski, who had recently become French citizens, also went into the Army and were assigned to the lab. They endured gas drills, blacked-out windows, and the absurdity of being told that now, what they were doing in their lab was considered a military secret. On October 30, 1939, Joliot deposited a sealed document with the Académie des Sciences; signed by himself, Halban, and Kowarski, it outlined the possibilities for power plants and bombs inherent in producing an "unlimited chain reaction."

With Poland vanquished and the major belligerents refusing to engage land forces, the conflict entered a period already being labeled as "the phony war," "*der Sitzkrieg*," or "*la drôle de guerre*." Perhaps as a signal to Great Britain of Germany's willingness to begin peace negotiations, Hitler issued an order stopping all sorts of long-range weapons projects and other military-related research. This decision was to have many consequences for Germany. Among the projects delayed were jet aircraft, better submarines, and more effective radar.

Also affected was German nuclear-physics research. A high-level nuclear-science committee had been formed, similar to that of the

United States, with Hahn, Gerlach, and Heisenberg as senior members, but without direct access to the head of state and also lacking in other essential attributes. As Michael Eckert suggests in his study of the German nuclear-bomb project,

> Scattered over many separate institutes, this project never had a coherent research strategy, nor was Heisenberg, its scientific leader, in total control of it. There was no teamwork between various institutes, and within a single institute hierarchical order and traditional barriers precluded useful collaboration.

In the fall of 1939 Heisenberg suggested to his scientific colleagues that while the rare uranium-235 could generate energy and heavy water could be used to brake and control the reaction, use of the more readily available uranium-238 was more promising, but it would require enrichment—an idea replete with technical problems whose solutions would be quite expensive. Since the war was going well for Germany, the government decided against funding Heisenberg's proposal to enrich uranium-238. None of the scientists fought this decision, nor did they deeply question their assumptions about what might be necessary to make a bomb—assumptions, as it turned out, that were fundamentally in error.

At Peenemünde, Hitler's slowdown and cutback order translated into reduced steel quotas that put a crimp in the rocket program just as it was having some success: at test launchings in October, rockets reached two miles into the sky before plummeting down on target areas in the sea. Citing that single success, Dornberger appealed to the army's commander in chief, Walther von Brauchitsch, to continue full funding of rocketry; von Brauchitsch did so without informing Hitler. Von Braun also assisted the cause by subcontracting with the universities and research institutes to solve specific technical problems. "In the polycratic ruling system of the Third Reich, no one was able to organize research centrally or produce a coherent scientific war effort," writes Michael Neufeld of the Smithsonian Institution, an expert on the history of rocketry; but, he contends, Dornberger and von Braun came close to making such a big-science project in rocketry by assiduously promoting "an academic atmosphere, including the exchange of ideas and research results among the Peenemünde laboratories and between the Baltic-coast facility and the universities." These collaborations produced better fuel mixtures, overcame engine problems, and designed the first fin-stabilized supersonic projectile, which artillery specialists had said was impossible. Soon the rocket group won a new convert, Albert Speer, Hitler's architect, who later wrote that he liked "these mathematical romantics" involved in "the planning of a miracle."

In October 1939 the Nobel Prize for physiology/medicine was awarded to German chemist Gerhard Domagk, then head of an I.G. Farben bacteriology laboratory. Before the war began, American, British, and French scientists had sponsored him for the Nobel Prize. He had discovered and cultivated a sulfonamide that held out the promise of curing one of the most prevalent causes of hospital and infant death, streptococcal infections. Based on his work, sulfapyridine was synthesized and became the drug of choice in treating pneumococcal pneumonias. That October he initially acknowledged the honor to the Caroline Institute in Stockholm but was then arrested by the Gestapo for doing so. Only after declining the prize was he permitted to leave prison and return to his lab.

Richard Kuhn also refused to accept a Nobel that year; the director of the KWI for medical research at Heidelberg was awarded the chemistry prize for his work on carotenoids and vitamins. In exchange for being permitted to remain in his laboratory, he began research in applying his theories to improving agriculture.

In early November 1939, while the "phony" phase of the war was still settling in, the British naval attaché in Oslo received a letter saying that if the British wanted information on German scientific and technical developments, they should alter the opening words of a daily news broadcast; they did, and the attaché promptly received seven handwritten pages and a small box. These were sent to London and thence into the hands of Reg Jones, the British physicist turned intelligence officer, who was in the process of completing his first report on potential German secret weapons.

The two letters thereafter called "the Oslo Report" astonished Jones. They cataloged the most technologically advanced weaponry then under development by the Third Reich; and the box, in Jones's words, "proved to be an electronic triggering device which, our correspondent said, had been developed so as to operate a proximity fuze in anti-aircraft shells." The identity of the author of the report was then unknown, but Jones concluded that it was someone quite familiar with a wide range of technically advanced projects: the letters described the progress of the dive-bomber, remote-controlled gliders, pilotless aircraft, rocket-propelled artillery shells, electric fuzes for bombs and shells, new kinds of magnetic torpedoes, aircraft range finders, flamethrowers, and smoke grenades, as well as the location of such secret military testing labs as Peenemünde. The report also stated that the British aircraft that had raided Wilhelmshaven had been detected "at a range of 120 kilometers by radar stations with an output of 20 kilowatts." It did not identify the wavelength but suggested that this could be determined and the transmissions jammed.

Jones believed that the report was accurate because of its technical detail and its mirroring of information that he had already gleaned from MI-6 debriefings of captured German soldiers and the reports of spies. His superiors, however, decided that the list of weapons had come too easily into British hands and was too fantastic, so that it must be a German plant; any one man, they contended, was unlikely to have such an encyclopedic knowledge of advanced technical matters as the letters demonstrated, something Reg Jones would have known had he not been so young and naive. They told Jones to stop working on the Oslo Report and threw away their copies of it. Independent-minded enough not to accept his superiors' judgment, Jones retained his copy and his respect for what it revealed about German science-based weapons; he also realized, for the first time, that the task ahead of Great Britain was daunting, since the country had very few answers for the weapons that the Germans were evidently readying for war.

Jones would not discover until almost thirty years later the identity of the author of the Oslo Report: Hans Ferdinand Mayer, the Siemens physicist. Disturbed by the torpedoing of the *Royal Oak* at Scapa Flow on October 14, he had first written an anonymous letter to the U.S. Embassy in Berlin, and when that received no response had arranged a business trip to Norway, there writing the letters that became the Oslo Report. In another missive he further fulfilled his obligation to Cobden Turner, suggesting that they maintain contact through meetings in still-neutral Denmark; to conceal his identity from anyone but Turner, Mayer signed the letter as "Martyl."

In his own laboratory, with the previous gift from Mayer, Cobden Turner began work on the forerunners of a device that would become the core of an important weapon, the proximity fuze.

The military labs in Great Britain were developing far-out weaponry that would be of only marginal use in the war. At Lindemann's request, Churchill commissioned Alwyn Crow, a rocket expert at the Projectile Development Establishment, to perfect a high-altitude minefield—the Lindemann idea so roundly denigrated by Tizard and others in prior years. These aerial mines were called "unrotated projectiles" or UPs, to contrast them to RPs, "rotating projectiles" shot from a barrel; eventually UPs were produced for all three British armed services. Another research unit, M.D.1, looked into "irregular" warfare devices such as plastic explosives and portable antitank weapons, and there were the also freelance investigations of maverick researcher General Millis Jefferis. On Churchill's behalf, Lindemann oversaw the experimental-weapons efforts and performed several other services for the First Lord: condensing reports, interpreting scientific and mathematical language, and producing endless charts that earned the encomium of being "things of beauty and ingenuity."

Lindemann's brother Charles, also an accomplished scientist, was then working with Tizard at the Air Ministry and decided to try to end the Tizard-Lindemann feud. Tizard was willing at least to bury the hatchet for the duration of the war—indeed, he and Prof had sat down together several times in the past few years, and Tizard had occasionally lunched with Churchill at the Admiralty. At the peace dinner the men formally shook hands, but Prof remained unwilling to resume a genuine friendship with Tizard.

The adviser to the chief of the Air Staff was keeping an eye on everything during the phony war: in addition to overseeing radar development in its several forms and maintaining the atomic-fission work, Tizard visited laboratories and gave assistance in the fields of gunnery, magnetic detection of submarines, better protection for bomber crews, map construction, and cockpit communications. He came to believe that only a half dozen innovations in weaponry or defense systems could be introduced during the course of a war and have a chance of materially affecting its outcome. His problem, and Great Britain's, was how to get the decision makers to agree on what those half dozen projects ought to be.

Across the Atlantic, Vannevar Bush was moving up. He became chairman of the National Advisory Committee on Aeronautics, while continuing to serve as president of the Carnegie Institution. He learned important lessons in politics from NACA executive director John Victory, who in backroom deals now managed to get an isolationist Congress to approve the Sunnyvale aeronautics laboratory, previously blocked by the appropriations committees, and to commission a third laboratory for military aircraft engines. On his own, Bush brought into NACA representatives from industry who had earlier been excluded, making NACA the agency best able to coordinate all of the country's aviation-related research. The difficulties involved in obtaining needed funding and authority for NACA to conduct research that was so clearly relevant to military needs contributed to Bush's feeling that had war come then to the United States, the country would have been sadly unprepared.

President Roosevelt was learning that the depth of American unpreparedness went well beyond the obvious matters of outmoded airplanes, submarines, and torpedoes, to the inadequacy of America's reserves of rubber, to the accumulated deficit in scientific research devoted to militarily related goals, and to the health of Americans of draft age. He had used the opening of hostilities as reason to accelerate the draft of young men into the Army, but word soon filtered back to him that physicians at intake centers were rejecting as unfit for service an inordinately large percentage of those they examined. Many potential draftees were in poor overall health, and a sizable fraction had never before been seen by a medical doctor; another group was rejected because they were toothless or required serious dental work; a third group couldn't see well enough and had no eyeglasses; a fourth group was illiterate, unable to read simple orders; and significant numbers of young men were psychologically impaired or unsuited for the discipline of the Army. The source of most physical and mental problems was the Depression. Roosevelt ordered a study to find out whether inexpensive and uncomplicated medical care—dentures, eyeglasses, vitamins, and better nutrition—could bring more young men up to the standard for induction.

In November 1939 Roosevelt declared the coastal waters off Great Britain, France, and Germany a war zone, which prohibited American ships from entering them. This action was part of the repeal of the Neutrality Act; another facet was the substitution of legislation enabling foreign countries to buy supplies from the United

States on a "cash and carry" basis. Such supplies were becoming increasingly important to Great Britain, because German mines were closing the southern ports. The British had to restrict shipping to a single lane in the Thames, but mines still sank 27 ships in November; the loss would have been larger had the Germans possessed more than 1,500 magnetic mines in their larder.

On the night of November 21–22, 1939, a British antiaircraft installation shot at a German mine-laying plane; the alarmed pilot loosed a parachute with a deadly package earlier than planned. The mine missed its target in the shipping lane, and landed instead on a mud flat near Shoeburyness. Word that it sat there, unexploded, reached Churchill, who asked that it be brought in and examined. At great personal risk, an officer and a civilian scientist retrieved it. Within eighteen hours the laboratory identified it as a TMB (Torpedo-Mine B) and deduced its mechanisms. Designed to float at a depth of fifteen fathoms, the TMB was the most formidable science-based weapon of the new war; it used a component of the earth's magnetic field as a detonator and was additionally, as a contemporary report put it, "full of beautiful technical tricks," such as having the clock that set its springs tuned to the latitude in which the mine was placed, and incorporating delay mechanisms and anti-mine-sweeping devices designed to explode it in the enemy's presence.

A Naval Convention signed in 1936 by Germany, Great Britain, France, and other countries decreed that if seagoing mines were laid, the country that deployed them was to publish their location so that surface ships could avoid those lanes. While the British and French published charts of the mines they laid to close the approaches to the German North Sea ports, the Germans, who used the darkest nights of the fall months to lay mines off the coasts of England, did not. (German mine-laying boats escaped detection because there was no British seagoing radar out looking for them.)

Within days of the recovery of the unexploded mine from Shoeburyness, a dozen engineers hired by the Admiralty had begun developing countermeasures. By Christmas Eve, 1939, the British were able to detonate a similar mine a hundred feet ahead of a minesweeper. Shortly thereafter they undertook a prodigious science-based task: using fluxmeters and current-carrying cables suspended around the hulls of ships or placed aboard to interrupt the magnetic field, they degaussed the entire fleet of the Royal Navy. The British cleared their southern shipping channels of mines, though not in time to prevent TMBs from heavily damaging the *Nelson* and the *Belfast.* Deconstruction of the German magnetic mine also helped the British fashion better magnetic mines of their own, which reinforced those already laid in the North Sea ports, contributing to a near-total blockade of Germany's northern waters.

Solving the magnetic-mine puzzle "was the first technical battle in which we won a decisive victory over the enemy; but more important still, it was one which brought science fully into the war in the very early days," Charles Goodeve later wrote; he was a Canadian physicist working with the British Navy.

In mid-December 1939, British warships cornered the *Admiral Graf Spee,* a German pocket battleship, and her captain scuttled her in the harbor at Montevideo. The *Graf Spee* did not sink completely, and the British recovered its radar and determined it was operating on a wavelength of 80 centimeters. This was evidence that German warships indeed had radar—something not previously confirmed by British Intelligence—and effective gun-laying radar at that, which the British did not possess. This important piece of news was reported promptly to the Admiralty, which managed to ignore the report for the next eighteen months.

After Germany and the USSR had carved up Poland, the Reich Education Ministry (REM) reversed a previous edict that had expressly forbidden contact with scientists of the Soviet Union. REM now asked German scientists what they wanted from the Soviets; the most frequently given answer was a surprise: information, in the form of scientific journals in English, French, and other languages that Germany was no longer able to obtain from the West. Although German researchers could not say so openly, without these journals they were falling behind in basic- and applied-research fields.

As for Soviet scientists, their commitment to the war effort did not change as a result of Soviet takeover of territory to the west of their previous borders, as Stalin believed that the USSR was in no immediate danger from Hitler, whose plans he believed involved France and the Low Countries. Nonetheless, Stalin began to relocate scientific institutes, factories, and agricultural production from the areas west of Moscow, sending them eastward, beyond the natural border of the Ural Mountains.

"Germanification" of the conquered parts of Poland, under the control of Heinrich Himmler of the SS, involved German scientists in setting up racial and psychological criteria for selecting new settlers, the breeding of plants, and the placing of Polish scientific institutes under the direction of the KWIs. The Berlin and Brandenburg KWIs were officially labeled important to the war effort, entitling them to raw materials for research at a moment when allocations for all nonessential organizations were reduced.

One scientist who came to the fore in the Germanification program was the Austrian animal psychologist Konrad Lorenz, a founder of the new discipline of ethology. Lorenz "never made a secret of his admiration for the new situation in Germany and the [Party's] accomplishment in all areas," a friend observed. When Germany took over Austria, Lorenz had joined the Nazi Party and became a member of its Office of Race Policy, receiving its permission to lecture on his comparative studies of instinctive behavior, which, Lorenz wrote, "We confidently venture to predict . . . will be fruitful for both theoretical as well as practical concerns of race policy." Lorenz began to espouse a worldview that "all physical and moral manifestations of decay that cause the decline of cultured peoples . . . are identical with the domestification manifestations in domesticated animals." From there it was only a short leap to concluding that "degenerative types" of humans, characterized by "coarser competitive methods," threatened to "pervade the *Volk* and the state and lead to their downfall," a process whose reversal could be achieved only by a "deliberate, scientifically founded race policy" of the sort being pursued in the newly conquered territories.

I.G. Farben's chief salesman, Baron Georg von Schnitzler, entered Poland close on the heels of the German troops, but his attempt to absorb the Polish chemical industry ran afoul of a general in the economics ministry. Aware of Farben's takeovers in Austria and Czechoslovakia the general allowed I.G. Farben to replace Jewish scientific technicians in Poland with Aryan employees but not to assume ownership of the Polish companies. Schnitzler went over the general's head to Göring, whom Hitler had asked to establish an organization to confiscate and dispose of Polish property in ways that furthered German war aims. Göring's attempt to help I.G. Farben was blocked by Himmler's deputy in charge in Poland. Schnitzler switched to cultivating the deputy and within months persuaded him to allow I.G. Farben to buy at inordinately low cost the entire chemical industry of Poland. Being beholden to Himmler came at a price: I.G. Farben's future in Poland would shortly include total involvement with the Auschwitz death camp, begun at the turn of the year in association with a synthetic-rubber plant. In scouting locations for that plant, chemist Otto Ambros—the company's leading authority on both buna rubber and poison gas—had to choose between a site in Norway and one in Poland; he chose the Polish one because the SS camp next door would provide a continuing supply of slave labor. With that in mind, I.G. petitioned the government for permission to also build a similar synthetic oil plant at Auschwitz.

Those plants would proudly bear the I.G. label, but another project was deliberately distanced from that label: a separate chemical company named Anorgana undertook construction of a factory at

Dyhernfurth, in western Poland, that would eventually cover more than a square mile and be devoted exclusively to the production of the Third Reich's most potent secret weapon, nerve gases. Back in 1937 I.G. Farben chemist Gerhard Schrader, trying to make a more effective insecticide, had found a very deadly compound he named tabun; within a year Schrader concocted an even deadlier gas, sarin. Dyhernfurth was to produce 3,000 tons of nerve gas a month; packed into artillery shells, one month's worth would be enough gas to wipe out the population of the British Isles.

Carl Bosch, still the nominal leader of I.G. Farben, was no longer talking to the firm's management, except to one old favorite. Bosch was sinking beneath the combined weight of physical decline, alcohol abuse, and depression brought on by a belief that his accomplishments—the syntheses of nitrates, fuel oil, and rubber—had brought Germany into a second world war. In the fall of 1939 he left Germany and moved to Sicily, taking along for company an ant colony borrowed from a Kaiser Wilhelm Institute.

In October 1939 the German government had issued an edict expanding the authority of physicians to order mercy killings; to emphasize that this was war-related, the edict had been backdated to September 1. In November or December of 1939 the "mercy killing" of children with severe birth defects began in Germany, followed by the first gassings of several thousand adult Germans certified as incurably insane in December 1939 or January 1940—the precise dates still cannot be pinpointed. The execution of people with severe insanity, mental retardation, or incurable birth defects had been permitted under the 1933 law. Objections from Catholics, which had previously prevented the program from being implemented, had now been eliminated, just as Hitler had told intimates it would be when war fever made it easier for the public to accept euthanasia. According to Franz Stangl, who later became an important bureaucrat in the euthanasia program, killings of patients initially took place only after examination by two physicians who administered four different tests to ascertain the incurable nature of the patients' conditions. Nonetheless, the killings were done in secret, and the program was run from an "inconspicuous villa" in the suburb of Berlin-Charlottenburg, the headquarters of the "General Foundation for Institutional Care."

The world's press began to report on the atrocities committed by the SS in Poland in the late fall and early winter of 1939–40, and to a lesser extent also reported on the euthanasia program. News of these Nazi excesses combined with tales told by German refugees to firm the resolve among the scientists of the Western countries to set aside their predilections for peace and use their expertise to defeat the Third Reich. In contrast to the emotional climate of World War I, when the Allied scientists' ruling passion had been patriotism, this time their motivation was the fervent belief that evil would triumph if good people did not actively combat it.

On November 30, 1939, the USSR bombed Helsinki, attacked Finnish islands from the sea, and sent troops over the border in an all-out attack similar to Hitler's blitzkrieg against Poland. But despite a huge numerical superiority, it took the Soviets several months to achieve the upper hand. Soviet military weakness in the Finnish campaign, and the lack of Soviet advanced weaponry, contributed to Hitler's belief that his forces could beat Stalin's.

Hitler had scheduled an invasion of the Low Countries for January 17, 1940, but on January 10 a Luftwaffe plane had to land behind neutral lines; its passengers included an officer carrying plans for the invasion, which were then presumed to have fallen into enemy hands. This loss, and information from deciphered Belgian and Dutch coded traffic that revealed how those countries would aid British and French forces in the event of a German incursion, caused Hitler to scrap the invasion plan. While his military chiefs scrambled to fashion a better way to invade the West, Hitler scheduled an invasion of neutral Norway, ostensibly to preclude the Soviets from taking over Norway after they vanquished Finland.

Later in January another plane went off course and crashed in the United States: the prototype of a bomber. The press soon discovered that one of the people on board was a French military observer who—over the objections of the Army Air Corps—had been given permission by President Roosevelt to view the plane. Isolationists in Congress seized on this incident to force a pledge from the president not to sell classified technology to other nations. A deal to trade information on America's Norden bombsight for information on equally valuable British equipment evaporated.

Around this time, in Great Britain, E. G. Bowen's radar research group completed the first air-to-surface set for use in planes searching the sea for enemy ships or surfaced submarines, but it was not accurate enough because no one had yet found a device to produce the ultrashort wavelengths that would provide clearer pictures. One such device was soon developed by two physicists at the University of Birmingham, John Randall and Henry Boot. The "cavity magnetron" was really a super power source, and one that was extremely compact. A prototype made in part from copper coins and held together by sealing wax was able to generate enough power not only to light a car headlight but to burn out the bulb in the headlight. The Randall and Boot device was the key to microwave radar that Bowen and everyone else had been looking for. The cavity magnetron almost instantly became one of the most prized and heavily guarded scientific secrets of the war.

In January 1940 cryptanalysts at Bletchley achieved a significant breakthrough of their own. As with most code breaks, it was a combination of good thinking, cooperation, and luck. Back in September, as hostilities in Poland wound down, French agents had managed a very important rescue, spiriting out of Poland to safety in France mathematician and code breaker Marian Rejewski, with several of his compatriots. Housed in a large villa in the suburbs of Paris, the Poles continued work on breaking the German codes, now alongside French and British cryptographers.

When the war had begun, the MI-6 unit at Bletchley had sent out telegrams to Oxford, Cambridge, and other universities, by prearrangement summoning dozens of mathematicians, champion chess players, crossword-puzzle whizzes, and German linguists to Bletchley Park to begin to solve "the Enigma problem." By December 1939, Turing and the cryptanalysts were coming closer to breaking the German codes by substituting German-language letters for numbers, but testing various substitutions was taking too much time. Turing helped devise an electric multiplier to make the work go more swiftly. The process reduced the number of combinations to be tested by means of frequency analysis (based on the frequency with which letters appeared), contact analysis (based on the frequency with which certain letters appeared next to others), and the knowledge that no letter was ever used as a code for itself while certain stock phrases were repeated word for word in many messages. Gradually individual messages were decoded; once a key had been broken, it would be used to adjust the settings on what were called "Bombes," relay-driven decoding machines that mimicked the workings of the Enigma, so that these could decipher parts of dozens or hundreds of messages recorded from German radio traffic. But it was still relatively slow going.

Once again it was Rejewski who provided the essential clue. Turing went to France to confer with him and returned with a sense of what the Brits had been doing wrong. Within days of resetting their Bombes, in mid-January 1940 the Bletchley code breakers deciphered the first complete bundle of Enigma messages, those transmitted between German army posts on a single day, two and a half months earlier. They still could not break the Luftwaffe's code or the much more difficult one used by the German Navy, which employed additional wheels on the Enigma machines. Two added rotors were captured in February 1940 from a U-boat sunk in the Clyde River.

These were not enough to allow the complete solution of the code, but they did prove that the Enigma was vulnerable. To ensure continued Allied decoding of future messages, a cloak of secrecy was thrown around the breakthrough; should the Germans learn of their careless mistakes, they could easily avoid making them.

The academic cryptanalysts had been summoned quickly to Bletchley, but an office culling the 7,000-man list compiled by the Royal Society to assign specialists to specific research units and tasks was placing only 50 to 100 people each week. The lackadaisical pace of British bureaucracy did not extend to aliens of German origin. Thousands were interrogated, had their backgrounds checked and movements monitored. Only the low level of hostilities during the winter kept the British from interning more aliens and from speeding up the assigning of scientific personnel to wartime tasks. Refugee scientists were also interned. When they were released, even if they wanted to work on the war effort, they were not considered for certain assignments, notably the development of radar. According to German-born physicist H. G. Kuhn, at the insistence of the Admiralty, anyone not born in the United Kingdom was forbidden to do radar work. But Kuhn and his émigré friends were permitted to use their talents to further the development of nuclear physics, since the authorities did not envision anything of practical use emerging from that study in the near future.

Early in 1940 the British and French made plans to send troops to Norway to cut off Germany's access to iron ore there and in Sweden and then to have the Allied troops cross to aid the Finns against the Russians. Hitler learned of the plan, scheduled for April, from decoded British messages. He readied his own preemptive strike against Norway, to begin just before the Allies started.

In advance of these scheduled invasions, an agent of the Deuxième Bureau, Lieutenant Jacques Allier, began preparations for an important trip to Norway. Frédéric Joliot had set in motion Allier's mission: to arrange for France to buy or borrow the entire stock of heavy water produced at the facilities of Norsk Hydro, which amounted to all of the heavy water then available in the world. During the previous few weeks, at France's request—and despite German ownership of 25 percent of Norsk Hydro—the company had been fending off demands from I.G. Farben for two tons of heavy water, citing as its reason the German firm's refusal to reveal why it needed so large a supply.

Allier traveled to Norway in early March, under his mother's maiden name, and taking other precautions to conceal the purpose of his trip; the Germans learned of it anyway, probably from deciphered code traffic, and tried to intercept him. Allier eluded them and on March 9, 1940, signed an agreement for Norsk Hydro to loan all of its heavy water to France for the duration of the war. The firm's general manager told Allier he would be resolute in keeping the heavy water out of German hands, even if he were to be shot for doing so.

Allier and three other agents carried twenty-six specially made canisters to the Oslo airport and by subterfuge spirited them to Scotland; another plane on which they had made reservations (but had not boarded) was forced from the air by German fighters and searched. The French agents then took the canisters through England—the British knew nothing of this—and across the Channel to Paris, where Joliot stored them in the cellars of the Collège de France. When, shortly, the Nazis invaded and took over Norway, they found only a depleted heavy-water supply at Norsk Hydro, and the lack of heavy water curtailed their experiments. It also deterred Japanese physicists from proceeding with an atomic-research program, because they believed that without heavy water they could not moderate chain reactions.

A few weeks before Allier's mission to Norway, two émigré nuclear scientists in Great Britain had an important conversation. As

Rudolf Peierls later recalled it, Otto Frisch asked him, in regard to what was needed to bring about fission, "What would happen if one had a pure uranium 235 in a sufficient quantity? How much would you need? And if you got it, what would happen?" To Peierls, "This seemed a rather academic question" but he had developed a formula to estimate such a thing, and "you could on the back of an envelope take my formula and find out what the critical size was." So he did, and was surprised to find it was much smaller than he had imagined. Then two émigrés asked themselves, "Now supposing you managed to set up a chain reaction, what would happen?" It took only a "couple of pages of calculations to show that you would get a very substantial fraction of the available energy released." They realized the importance of these calculations, and that "we might be the first ones to think about this [and] in a country at war it was our job to make that available to the right authorities." So they wrote a paper. Feeling that he could not entrust this material to a secretary, Peierls typed it himself and in mid-March 1940 sent it to Henry Tizard.

It was entitled "On the construction of a 'super-bomb' based on a nuclear chain reaction in uranium." The memo laid out in stark, clear language a blueprint for constructing an atomic bomb from an amount of enriched uranium that was breathtakingly small—just one pound, as opposed to the thousands of pounds previously thought necessary. Frisch and Peierls likened the force of an atomic bomb to "the explosion of 1,000 tons of dynamite" that would for an instant "produce a temperature comparable to that in the interior of the sun" and could "destroy life in a wide area." The German-born physicists were adamant that such a weapon could not be exploded anywhere without killing large numbers of civilians, which "may make it unsuitable as a weapon for use by this country," but they warned that "since all the theoretical data bearing on this problem are published, it is quite conceivable that Germany is, in fact, developing this weapon." As evidence they cited the development of isotope-separation tubes by a Munich professor of physical chemistry.

Emphasizing that since there was no way to protect against such a bomb, "the most effective reply [to Germany] would be a counter-threat with a similar bomb [and] therefore it seems to us important to start production as soon and as rapidly as possible, even if it is not intended to use the bomb."

Tizard had earlier given only modest support to the physicists involved in looking into the possibilities of making a bomb, but he now assented to suggestions that a committee be formed, under Thompson and including Blackett, Cockcroft, Chadwick, and Oliphant, to further examine the matter. At that committee's first meeting these men were addressed by Jacques Allier, who told them of securing the heavy water from Norway.

The April 1940 German and Allied invasions of Norway were nearly simultaneous, but German foreknowledge of Allied plans, based on deciphered radio messages, contributed to a victory by the Third Reich that also brought them control of Denmark.

During the period when Germany and the Allies were just landing in Norway, Carl Bosch left Sicily and returned to Germany. His physical and mental condition had deteriorated further, and he seemed to know he was dying. He told his doctor that Hitler was certain to win in Norway and would then invade and defeat France; he also prophesied that Hitler's victories would not last, because the Führer's mental illness would eventually result in the destruction of two entities that Bosch loved with equal fervor, Germany and I. G. Farben. Bosch died on April 26, 1940, at the age of sixty-five.

Despite losing the ground war in Norway, the Allied navies and air forces destroyed all ten of Germany's cruisers and captured from a trawler a sheaf of cipher documents that provided a key to solving the German naval Enigma. As of April 15, 1940, the Bletchley facility became able to decipher German radio messages almost as quickly as they were transmitted. However, because there were so many messages and because the radioed ones were only a fraction of the total German communications traffic, which also included teleprinter messages, telephone calls, and hand-carried notes—to which the British had no access—Bletchley was not always able to know precisely what the German forces were up to.

A personal message received in Great Britain just then also caused some consternation. Sent from Lise Meitner in Sweden, it read, "Met Niels and Margherita recently. Both well but unhappy about events. Please inform Cockcroft and Maud Ray Kent." In the wartime atmosphere, this message was thought to have cryptic meaning, especially since no one in the scientific community knew to whom "Maud Ray Kent" might refer. A one-letter substitution and some juggling around made that phrase read "Radium Taken," while another scrambling suggested it might be "Make Ur Day Nt," suggesting an imperative to hasten British atomic-bomb research. It was sufficiently intriguing that the Thompson committee, charged by Tizard to determine whether a bomb could be made, now decided they should be called the MAUD Committee, and they speeded up their efforts.

On May 7, 1940, as Hitler's forces massed at the borders of Belgium, the Netherlands, Luxembourg, and France, a debate raged in the British House of Commons over the allies' inability to defend Norway. The downfall of the Chamberlain government followed the debate, and on May 9 parliamentary leaders agonized between choosing the foreign secretary, Lord Halifax, or Winston Churchill as the next prime minister. At the same moment, in Paris, Paul Reynaud offered his resignation as premier because his cabinet refused to dismiss Maurice Gamelin, the commander in chief who Reynaud believed had not readied France for a German onslaught.

On the morning of May 10, 1940, the Third Reich's forces attacked the Low Countries and prepared to enter France through the Ardennes Forest. Reynaud and Gamelin retracted their resignations. During his first cabinet meeting as prime minister–designate, Churchill showed off the latest scientific gadget, a prototype of a homing antiaircraft fuze.

The new German invasion strategy, devised after the earlier plans were presumed lost in the January plane debacle, was brilliant and unexpected. Like a matador's cloak, it drew Allied forces north toward Belgium and Holland, far enough away from the Ardennes to be unable to return south in time to counter the sword thrust of German tanks and other mobilized forces stabbing through a narrow opening into the French heartland. By the time Churchill as prime minister was pledging to the Commons "blood, toil, tears, and sweat" in the pursuit of "victory at all costs," the cloak-and-sword maneuver had broken through the French lines. As British and French forces tried to recover and prevent the Germans from piercing farther into France, Nazi troops completed overwhelming Belgium, the Netherlands, and Luxembourg.

In the midst of the German invasion of the Netherlands, which produced widespread panic, the Dutch government ordered its two radar experts to flee to Great Britain along with whatever equipment they could carry; they worked thereafter with the Allies, one on radar-jamming equipment at the British Signals school, the other in the Far East, fitting radar sets onto Dutch submarines that had been in the East Indies and so were not captured by the Germans.

The Germans did seize most of the Dutch submarine fleet intact in its home base; the vessels contained a unique feature, a "snorkel" that allowed the diesel-powered submarines to remain submerged at periscope depth while recharging their batteries with needed air. Although the snorkels would have been immediately useful to the U-boat fleet, they were disparaged as not of German origin and removed from the captured submarines before those vessels were recommissioned in Dönitz's slowly growing fleet.

By May 20, 10 panzer divisions had trapped 60 Allied divisions between the Channel and the Somme. Hitler savored the moment, imagining capitulation by the French and a peace treaty with Great Britain in which Germany would regain her former African colonies—while he failed to issue the order that would have resulted in capture or annihilation of the trapped Allied divisions.

On May 24, 1940, in Washington, D.C., the headlines told of German forces held only up by fog from attacking the British then gathering at Dunkirk. The final meeting of the Committee on Scientific Aids to Learning was held that day at a lunch at the Cosmos Club, and it turned into a discussion about American scientific preparedness for the war among Vannevar Bush, Harvard president James B. Conant, Bell Labs leader Frank Jewett, and a few others.

Since January, Bush had become increasingly frustrated at his inability to obtain greater cooperation in war preparedness among government, military, private industry, and academic research institutions. At Bush's urging, in February the Air Corps asked the National Academy of Sciences (NAS) for advice on defense against new aircraft; the NAS convened a committee chaired by a Caltech professor that had leisurely held a few meetings and produced nothing of value to the Air Corps. This signal failure helped convince Bush that a new agency was imperative, because the current system "would never fully produce the new instrumentalities which we would certainly need, and which were possible because of the state of science as it then stood."

Bush found allies in Hap Arnold, chief of the Army Air Corps—who, unlike other military men, believed in the power of science and new technology—and in Jewett, who had been instrumental in obtaining the Carnegie Institution post for Bush; in return, Bush had assisted Jewett's recent bid to become president of the National Research Council. Bush also enlisted Conant, Compton, and the independent researcher Alfred Loomis. A retired Wall Street lawyer who generously funded the research of Ernest

Lawrence and other innovators, as well his own investigations, Loomis was also a cousin of Henry Stimson, a Republican stalwart who had served in the cabinet during the Great War and again under Hoover.

Any one of these men could have become a credible leader of the American scientific establishment during World War II, but it was Bush who invented the position and the organization to bring the power of American science to bear on military matters. At the lunch he did not tell his companions that he had already asked Roosevelt's uncle, Frederic Delano, chairman of the National Resources Planning Board, to set up a meeting for him at the White House.

Though the luncheon companions were convinced that, for the good of the country, American science must be mobilized on military-related projects, others disagreed. There were two dangers. At one extreme, and arguing from what had occurred in previous wars, there was the likelihood that whatever the laboratories now came up with would not affect the current war, and therefore that research—all research, in universities and in military labs—should be shut down and the money saved put into procurement of already-developed weapons systems. At the other end of the spectrum, and arguing from the successful development of poison gases in the Great War, there was the possibility that warfare might become exclusively a matter of science-based weapons and counterweapons, which would put scientists in the position of making decisions for the military, even of eclipsing the military's authority to defend the country. Bush had to find a middle path: a way of using scientists for the national defense that would help rather than upset the military and that would preserve the military's ability and authority to conduct war.

On May 25, with the fate of the British Expeditionary Force hanging in the balance, Roosevelt petitioned Congress for an Office of Emergency Management to facilitate production for the military and for a substantial increase in the military budget, including provision for constructing 50,000 planes in the coming year—the past year only 2,000 had been produced. That day his secretary received Delano's request about Bush.

While Bush waited for an answer from the White House, the situation of the democracies in Europe continued to deteriorate. During the rest of May and the first week of June, as the British evacuated 200,000 of their own and 100,000 French troops and were also forced to escape from Norway, Germany consolidated control of the northern half of France and of the Low Countries. In the skies the RAF was losing three planes for every two of the Luftwaffe it shot down. Each night between 100 and 200 German planes bombed the British Isles, and the British believed that an invasion might be imminent. A study done for the War Cabinet concluded that the Royal Navy could not stop an invasion of Great Britain and that, once landed, the Germans might well fight their way to London. Churchill ordered tanks of poison gas loaded onto British planes, to be sprayed on German forces if they landed on British soil. Tizard suggested sowing the beaches with lightbulbs filled with mustard gas, an idea seconded by the gas-testing facility at Porton Downs. A million Molotov cocktails were also readied.

Around this time, as well, President Franklin D. Roosevelt made up his mind to run in the fall for a third term as president. He also asked Congress for an additional $1.7 billion to expand the Army to 375,000 men, rebuild the Navy, and purchase 50,000 planes. Some 600 freight cars filled with artillery guns and rifles sat on the docks, awaiting shipment to France, while Army bureaucrats, abetted by congressional intransigence, held up their transfer to an embattled France. During the first week in June, Roosevelt's closest aide, Harry Hopkins, met with Vannevar Bush. Though the men were philosophically at odds, biographer Robert Sherwood later wrote, they were of similar temperament: like Hopkins, Bush was "thin, quick, sharp and untrammeled in his thinking. He knew what he was talking about and he stated it with brevity and, like Hopkins, a good sprinkling of salt." Bush's philosophy for the new organization he imagined was based on a belief, which he articulated in a speech, that "know-how" could be "acquired only by constant research" and that the quantity of research was as important as the quality in obtaining quick answers to questions about such things as the design of new aircraft.

On June 10, as Paris was about to be overrun, Mussolini declared Italy to be at war with France; Roosevelt memorably charged that "the hand that held the dagger has struck it into the back of its neighbor." Aides had warned Roosevelt that this phrase might cost him the Italian vote in the next election, but the president had insisted on using it. On June 12 Bush and Hopkins met Roosevelt in the White House. Bush presented a one-page document recommending a National Defense Research Committee (NDRC) composed of military chiefs, the secretaries of War, Navy, and Commerce, and members from the National Academy of Sciences, and chaired by himself, with adequate funds to "correlate and support scientific research on mechanisms and devices of warfare." Bush anticipated tough questioning, but there was none; after fifteen minutes, Roosevelt signed "O.K.—FDR" on the paper and began the marshaling of the world's largest corps of scientific experts. A survey would eventually reveal its dimensions: 6,800 physicists, 60,000 chemists, 3,400 chemical engineers, 57,800 electrical engineers, 2,750 radio engineers, and 55,000 mathematicians.

Within days Bush gathered his core civilian leaders: Compton, Jewett, and Conant, supplemented by Caltech physicist Richard Tolman as a West Coast representative. Conant wanted the NDRC to build laboratories and hire staff; Bush rejected that idea in favor of awarding contracts for research to existing universities, research institutions, and industrial labs. "I shall never forget my surprise at hearing about this revolutionary scheme" of awarding contracts to laboratories to do research for the government, Conant wrote: he recognized that this profound change would bring about a new alignment of the scientific power structure. Bush saw the establishment of the NDRC as "an end run, a grab by which a small company of scientists and engineers . . . got hold of the authority and money for the program of developing new weapons." It was also a way to force the military establishment to pay attention to whatever that small company of scientists and engineers would develop.

The government of France declared Paris an open city on June 14, 1940, and the Germans swarmed in. On June 16, at a retreat 160 miles south of Paris, near Orléans, the French cabinet voted to seek an armistice.

That afternoon, Lieutenant Jacques Allier, the Deuxième Bureau Agent who had spirited the heavy water out of Norway, arrived on the doorstep of the Joliot-Curies in Clermont-Farrand. Joliot was determined to remain in France, in part because Irène was not well, in part to keep the laboratory functioning, but they planned for Halban and Kowarski to take the heavy water to England and continue research on nuclear fission. That evening Joliot predicted to Kowarski the entire course of the first part of the war: France would be only partially occupied; then Germany would try to subdue England from the air and fail to do so, or to mount an invasion; for a while England alone would resist Germany, until Russia and the United States would enter the war.

Next morning Halban, Kowarski, their families, and the heavy water went to Bordeaux and boarded a British collier under the command of a tattooed man stripped to the waist. The piratical-looking man was the twentieth earl of Suffolk, on a mission to carry to England machine tools, industrial diamonds, the heavy water, and a dozen scientists. German planes were already bombarding the harbor. Joliot's associates and the heavy water made it to England, but as Kowarski later recalled, "Nobody knew what to do with us. The government didn't know whether we were a valuable addition to the wartime effort, or refugees to care for, or potential spies. . . . We never knew whether we were honored guests or war prisoners." Kowarski told his wife that they had escaped the German occupation of Europe, yet carried that German occupation with them.

| CHAPTER FIVE |

Battles Above Britain

On the morning of June 21, 1940—two days after German planes first dropped bombs in the vicinity of London and a day before France accepted Hitler's terms for surrender—twenty-eight-year-old Reg Jones received a note to report immediately to the Cabinet room at 10 Downing Street. He thought it a practical joke but decided to answer the summons. In the Cabinet room with its long table, embossed chairs, and fireplace, he found Prime Minister Churchill, flanked by Lindemann and Lord Beaverbrook, together with Tizard, Watson-Watt, and the heads of the Air Force. "The atmosphere was clearly tense and perhaps even that of a confrontation," he later wrote. After listening to a few minutes of discussion of the possible existence of German radar, Jones felt that no one in the room knew more than he did, and he asked Churchill, "Would it help, sir, if I told you the story right from the start?"

"Well, yes it would," the prime minister said, and Jones held the Cabinet room spellbound—and Churchill speechless—for twenty minutes as he traced what had been learned about Knickbein from the evidence of downed planes, overheard prisoner conversations, interrogations, Enigma intercepts, and aerial reconnaissance. Churchill later wrote that the story that emerged from this "chain of circumstantial evidence, for its convincing fascination was never surpassed by the tales of Sherlock Holmes or Monsieur Lecoq." Knickbein, whose literal meaning was "crooked leg," was a radio-beam system for guiding a bomber to its target. One beam, 400 yards wide, set the course for the plane; another beam intersected the first at a designated distance from the target, thereby telling the navigator when and where to release the bombs; the information came into the planes through the Lorenz blind-approach system used to land planes during rough weather. The existence of Knickbein implied an ordeal for Great Britain, which had no way of combating night bombing attacks.

In an instance of a good scientist being at a loss because his knowledge was out of date, Tizard discounted the possibility that Knickbein could direct bombers, citing a long-held notion that radio waves would not bend sufficiently around the curvature of the earth to reach from Germany to Great Britain. Jones said that he had recently found an expert whose new data refuted that notion, and a few days earlier had used those data to quell Lindemann's similar objections.

Churchill then asked what was to be done about the beams. Jones suggested confirming their existence from the air, then jamming them or projecting false beams to make bombers drop their payloads early. Churchill agreed but also seized the moment to advocate Lindemann's aerial mines, which might be particularly effective against German bombers flying along beams; banging the table angrily, the prime minister reminded everyone that he had been pressing this case for years, but, "All I get from the Air Ministry is files, files, files!"

Shortly he would allocate £1 million to the design and production of the mines.

As for Knickbein, "That very day in June," Churchill later wrote, "I gave all the necessary orders . . . for all counter-measures to receive absolute priority. The slightest reluctance or deviation in carrying out this policy was to be reported to me. . . . This, however, was not necessary." Perhaps Churchill never learned that the following day a high Air Ministry official attempted to block a flight to confirm the existence of Knickbein; Jones threatened to tell Churchill. The flight was made, and its observations ratified every part of Jones's guesses. British technical establishments started Project Headache, finding ways to make "meacons," or masking beacons, to disrupt any Knickbein-directed offensive. One additional result of the June 21 meeting: because it implied that the developers of British radar had failed to recognize an aspect of radar from which the Germans had constructed a tremendous threat, Tizard offered to resign as adviser to the chief of Air Staff.

Tizard's resignation was not formally made and was soon brushed aside. But Tizard and the members of his prewar aerial-defense committee continued to bristle because now, among the scientists, only Lindemann had the prime minister's ear. Lindemann's counsel was as often inept or bizarre as it was savvy; for instance, he championed setting fire to enemy crops by means of dropping incendiary pellets, sandwiches of two layers of cellulose with a layer of phosphorus in between. A. V. Hill noted to his military superiors that such a device was bound to fail technically but, more important, that it was morally indefensible and would "give the enemy a strong hand for propaganda, and the excuse for stealing the crops and stores of our friends in occupied territory." The incendiary-pellet idea was put aside for a while, but the Tizard bunch could not prevent the spending of the million pounds for the aerial mines, even though Hill prepared another memo with fourteen detailed instances of the "waste of time, money and effort which [Lindemann's] advice has caused and of the improper methods by which he has worked."

In mid-June 1940 Stalin took the opportunity of Hitler's invasion of the West to occupy Latvia, Lithuania, Estonia, and half of Romania. Hitler fumed that these annexations had not been part of the Nazi-Soviet accord, but he could not counter them, focused as he was on consolidating his hold over Northern Europe. He also refused to launch an all-out air attack and invasion of Great Britain. Rather, flush with victory, he ordered time-consuming preparations for conquering the British Isles. Admiral Dönitz, chief of the U-boat service, advanced his headquarters to the Brittany coast and found that locating U-boat repair bases there saved a week per sortie and enabled better use of smaller submarines, because the distances to be covered were not as great as when the U-boats had been based in Germany. Before the fall of France there were 2.35 U-boats undergoing repair for every 1 in action; after the Germans took over the French ports, the figure fell to 1.84 U-boats in repair for every 1 in action, a 22 percent improvement in availability. Part of that improvement came from capturing France's good supply of copper, needed in the repair of U-boats.

The British anticipated and precisely imagined an imminent German invasion: a 5,000-man parachute force seizing seven key British airfields, while German bombers, fighters, and naval guns prepared the way for a 20,000-man assault on the British beaches. The War Cabinet placed eight army divisions near the beaches; had a huge tank ditch built; floated mines attached to scaffolding to prevent landings; demolished the centers of pleasure piers so they could not be used as jetties; and had pipes laid in back of the beaches so the sand could be sprayed with gasoline and ignited. One squadron of British planes was loaded with smoke-producing and poison gas as an ultimate defense against invading troops.

Over-influenced by years of German propaganda, the British Air Ministry initially estimated that the Luftwaffe had 2,500 bombers capable of dropping 4,800 tons of bombs daily. Lindemann thought these figures unrealistic and pressured the ministry to back them up. This resulted in revised figures of 1,250 German bombers capable of dropping 1,800 tons of bombs daily, assessments soon verified by Enigma decrypts, which served to give the RAF confidence that it was not overmatched. Later in the year an independent inquiry commissioned by Churchill put the relative strengths of the Luftwaffe and RAF even lower, at four to three.[1]

In Paris, Joliet was interrogated by two German physicists not well regarded within the scientific community but who because of their political views had been appointed to lead institutes. They were accompanied by Walter Bothe, who knew what he was doing, and by Wolfgang Gentner, Joliot's former assistant, who had worked on a cyclotron at Berkeley and was then building a cyclotron in Heidelberg. In front of his German superiors, Gentner was tough with Joliot, trying to find out what had happened to the heavy water and to the uranium ore. Joliot told the Germans the heavy water had left Bordeaux by boat, but he gave them the name of a ship that he knew had been sunk. He also knew that the ore had been hidden near Toulouse but claimed ignorance, and no one could gainsay him. Later, in a private meeting at a Left Bank bistro, Gentner and Joliot concocted a plan for Gentner to supervise the completion of the cyclotron in Paris, so it would not be shipped to Germany, in exchange for Joliot's remaining as director and concentrating on nonmilitary applications.[2] Bothe ratified the arrangement, agreeing with Gentner that Joliot's cyclotron could not be reestablished in Germany in time to affect the end of the war, which was expected

[1]The four-to-three ratio was confirmed by postwar tallies, based on documentary evidence.

[2]Joliot would verify that the Germans were doing only nonmilitary research by secretly visiting their laboratories at night.

soon. The Gentner deal gratified the German hierarchy, which could point to a French scientist of Joliot's stature working under their direction without complaint.

The Germans now had at their command the world's only heavy-water factory and thousands of tons of uranium ore in Belgium and access to more in the Congo; they had lacked only a working cyclotron to measure nuclear constants. Now they had one, and with it, all that was necessary physically to produce an atomic bomb.

On July 10, 1940, stepped-up German aerial bombing of Great Britain began, and on July 16 Hitler issued orders to proceed with preparations for an invasion: two weeks of probing by the Luftwaffe would determine the extent of Great Britain's defenses, and six additional weeks of bombing would soften up the British Isles for a September crossing. On July 21 Hitler told his military commanders to prepare also for an invasion of Russia; recently Stalin had ordered the formal incorporation of Lithuania, Latvia, and Estonia into the USSR, and Hitler believed he must stop the Communists from coming closer to Greater Germany.

Hitler's plan to subdue Great Britain's air forces rested on several mistaken assumptions. Among the pieces of equipment left behind by the British near Dunkirk was an early British radar set. Churchill had been quite upset at the prospect that the Germans might find it. Located and investigated by German technicians, it was pronounced an inferior copy of a design that the Germans had already eclipsed, something that could not possibly deflect the Luftwaffe from its task over Great Britain. Although information about British radar was known to German Intelligence, the division that produced the report of the RAF's strength on which Hitler relied made no mention of radar defenses. The Abwehr also underestimated the readiness of the British air fleet and ground and coastal defenses, and misread the capacity of British factories to turn out replacement planes.

During the heat of the Battle of Britain, and ever since, great emphasis has been placed on the role of those whom Churchill memorably labeled "the few," the pilots of the RAF, in saving the British Isles from destruction and subjugation. Their heroism is beyond doubt, as is the terrible regularity with which a significant fraction of pilots in each squadron were killed during nearly every sortie. Yet while the pilots' role in the survival of Great Britain was critical, so was that of the Home Chain of radar stations (operated primarily by women) and of the communications system that relayed their information about attacking planes to the RAF and ground defenses. A Spitfire required thirteen minutes to scramble and climb to 20,000 feet to engage the enemy; radar gave RAF squadrons in the south and west twenty minutes' notice of Luftwaffe attacks, and the extra seven minutes allowed defending squadrons to get into position to disrupt an incoming formation. Radar is considered to have doubled the efficiency of British fighter forces. Adolf Galland, the German ace who rose to command of a Luftwaffe division during the Battle of Britain, later wrote,

> The British had from the first an extraordinary advantage . . . their radar and fighter-control network. It was for us and our leadership a freely expressed surprise, and at that a very bitter one, that Britain had at its disposal a close-meshed radar system . . . which supplied the British Fighter Command with the most complete basis for direction imaginable.

Blindness toward the power of a technology that Germany's own scientists could have explained prevented Hitler and Göring from obliterating the Home Chain. Had they done so, they would have crippled the RAF response and achieved the Luftwaffe daytime air superiority that Hitler considered a prerequisite for invading the British Isles. It took the Germans most of the summer to revise tactics based on the erroneous belief that RAF fighters were tied operationally to particular airfields and therefore limited in mobility, and to realize that British planes were being sent to specific targets in the air by a sophisticated information and communications system. During the first part of August, Göring did bomb Home Chain stations, causing some damage, but in a conference at Karinhall on August 15 he told subordinates there was no point in further bombing the radar stations because "not one of those attacked has so far been put out of action."

"You are about to land at dead of night in a rubber raft on a German-held coast. Your mission is to destroy a vital enemy wireless installation that is defended by armed guards, dogs, and searchlights. You can have any one weapon you can imagine. Describe that weapon."

Vannevar Bush began putting this question to his recruits for National Defense Research Council work in the summer of 1940. If the question was designed to start the individual scientist thinking in new ways, it also reflected his conviction that since the professional military was unlikely to be the source of new weaponry that could win the war, civilian scientists putting their minds to military problems would have to solve them. The president's uncle, Frederic Delano, and his resources board were in the midst of compiling a National Register of Scientific and Specialized Personnel—akin to the British list but readied more quickly—in association with the NDRC. From his offices at the Carnegie Institution, Bush began recruiting scientists and awarding contracts.

Everything did not go smoothly. Bush's associate Merle Tuve, the mercurial head of the Carnegie Institution's magnetism laboratory, later recalled his frustration at that period: All spring, he had wanted to work on explosive devices, feeling that to continue on with his regular work was meaningless and that he must use his talents in "stopping this conflagration." But the military took two more months to clear him for work on what was then called the "influence fuze." On the other hand, retired Wall Street lawyer and amateur physicist Alfred Loomis, tapped to lead an NDRC committee looking into the possibilities of microwave radar, was quickly able to bring together Bush's former colleague (and sometime foe) Edward Bowles and Rochester physicist Lee A. Dubridge to start a massive electronics-research effort at MIT.

It was at this point in time that the British completed preparations to send a mission to the United States, bearing the scientific equivalent of the crown jewels.

Despite Bush's having already put together an impressive array of scientists by the time the British scientific delegation arrived in the United States, the only definite appointments that it had on its calendar were with the American military. None had been scheduled with Bush or with the NDRC.

Although Churchill and Roosevelt began trading messages in September 1939, and since then there had been several thrusts at Anglo-American cooperation, both sides were reluctant to share scientific military information. The great American secret was the Norden bombsight, developed for the Navy, which the British had been desperate to obtain. Their appetite had increased since April 1940, when British observers had been present at a demonstration of the bombsight at Fort Benning, Georgia. More than a dozen bombers all hit targets resembling the deck of a battleship with great accuracy from heights of 10,000 to 12,000 feet. The British plied American officers with alcohol and attempted to commit espionage to acquire a bombsight or knowledge of its mechanisms but obtained nothing.

After the fall of France, Roosevelt became more willing to listen to ideas for a scientific exchange and equipment trades with Great Britain. Then Churchill tried to kill it. The prime minister wrote a note wondering, "What is the urgency of this matter?"

> Are we going to throw our secrets into the American lap, and see what they give us in exchange? . . . I am not in a hurry to give our

> secrets until the United States is much nearer to the war than she is now. I expect that anything given to the United States services, in which there is necessarily so many Germans, goes pretty quickly to Berlin in times of peace.

Cooler heads prevailed, especially that of Lord Halifax, the foreign secretary, who insisted that general improvement in Anglo-American relations depended on "the habitual exercise of friendly assistance freely given, together with a general recognition of the other country's needs and feelings." Churchill then assented to the mission—and to having it led by Sir Henry Tizard.

Tizard's appointment was an acknowledgment of his value as a leader of scientists but was also an attempt to remove him from London and thereby prevent him from second-guessing Lindemann in the midst of Great Britain's most serious military crisis. For Tizard, the mission was a continued lifeline to the War Cabinet's prosecution of the war, at just the moment when the benefits of his decade's worth of effort were being felt but when his influence was being eclipsed. On August 1 Churchill again withdrew permission for the Tizard Mission, citing Roosevelt's refusal to have the United States manufacture British tanks in American factories; but within a few days Churchill was convinced that canceling the mission now would do more harm than sending it.

Tizard's team included A. V. Hill, Cockcroft, Bowen, R. H. Fowler, and other men of high scientific standing and considerable technical expertise, plus senior service officers. Chief among the Tizard Mission's crown jewels was the cavity magnetron, which could provide much greater power and enable the production of much smaller wavelengths than existing radar. Sets incorporating cavity magnetrons, it was hoped, could detect targets as small as submarine conning towers, make better range calculations for antiaircraft guns, and—because the sets would be so tiny—they could more readily be mounted in aircraft to help them fly effectively at night or in bad weather. Nonetheless, for more than a month the British cavity magnetron lay hidden in a box under a secretary's bed in a Washington hotel room, almost forgotten, while the Tizard Mission held meetings with the American military services.

The American brass was cool to the exchange, on two counts. They were concerned that Great Britain could lose the war and with it any American secrets that had been conveyed. They also feared that the exchange would lead to more NDRC contracts with universities, which the military still considered overly pacifist centers having a high potential for leaking secrets. In meetings with Tizard the Navy again refused to consider handing over the Norden bombsight, though the Army Air Corps did offer a Sperry bombsight touted as almost as good. The biggest response Tizard got out of the American military was when he showed documentary film of air battles, including footage of blips appearing on radar screens and of on-board cameras showing British fighters as they shot down German planes identified by radar. The film was evidence that what the Brits were saying was based on field experience; it was also a healthy antidote to cables sent from London by Ambassador Joseph P. Kennedy, who predicted the imminent demise of the British Isles from the Luftwaffe onslaught.

Tizard worried about failure of the mission until he and Fowler had a private dinner at the Cosmos Club with Bush and Tolman. The four men found they shared the belief that civilian scientists could and should influence the direction of military research and priorities. Bush and Tizard were very much alike—politically conservative, knowledgeable, practical, gregarious among their peers, good managers—though the American was far more ambitious and intent on building an empire. After this dinner real cooperation commenced, though Bush still had to obtain permission from the Navy for the NDRC to work with Tizard. The working scientists met, too: Cockcroft and Bowen with Loomis, whose group was already looking into microwave radar and therefore recognized the revolutionary potential of the cavity magnetron. Another meeting of this group, which included Bowles and took place at Loomis's estate in Tuxedo Park, laid the foundations for the American manufacture of the magnetrons by Bell Labs and Raytheon. Henry Stimson, Roosevelt's secretary of war, noted in his diary his cousin Loomis's excitement at the magnetron and his claim that "we were getting the chance to start now two years ahead of where we were."

Stimson also jotted in his diary Loomis's belief that "we were getting infinitely more from the British than we could give them." This was true and a source of annoyance for the British, who would return home with not much booty, because, as Tizard explained to Churchill, "Broadly speaking, the Americans are far behind us at present in technical equipment for war." But Tizard understood after his visit to the United States, and the British Cabinet would eventually acknowledge, that what Great Britain gained in the exchange was the certainty that the innovations they brought to Washington would be developed in the United States to the point of manufacture, then produced in the quantities and at the speed that would permit the innovations to become decisive factors in the war. Such assurance of manufacture could not then be obtained in Great Britain, for a complex of reasons including poor interaction between industrial companies and the government, interservice rivalries, and the dearth of raw materials.

Tizard and his mission performed an even greater service to both countries by helping to make the United States into an ally and the major resource for Great Britain more than a year before the attack on Pearl Harbor and by establishing permanent links among American, British, and Canadian (ABC) research-and-development efforts in many fields, including radar, fuzes, antisubmarine warfare, airplane design, chemical and biological agents, penicillin, and, eventually, nuclear fission and a potential atomic bomb. Tizard himself was a key to this cooperation; he fought sniping at home from the British brass, who wanted Cockcroft and Bowen to return home. Tizard was able to convince Churchill that these valuable working scientists must remain in North America to direct American and Canadian scientific efforts—the United States and Canada had just signed an agreement to establish a Permanent Joint Board on Defense of the northern half of the Western Hemisphere—and to make arrangements there to manufacture equipment that would be militarily useful to the British in the immediate future.

Cockcroft and Bowen proved to be of tremendous value in the North American war efforts. The Welshman Bowen, given the nickname of "Taffy" by his MIT colleagues, advanced American radar work in many ways. Cockcroft advised both American and Canadian nuclear researchers, and later during the war took the senior management position in the Canadian part of the building of an atomic bomb. In 1951 Cockcroft would share the Nobel Prize with Ernest Walton for work they had done jointly in 1932 at Rutherford's laboratory. But while Cockcroft was an important player on the scientific side of World War II, Walton—of the same generation, training, and interests—essentially sat out the war in his own laboratory and taught at Trinity College in Dublin.

In August 1940, Jacques Allier was summoned to report to Pierre Laval, Admiral Darlan, and others in the Vichy government on what had happened to France's nuclear-research program. Laval expressed regret that France's heavy water had been sent to Great Britain, as he would have preferred to turn it over to Germany, but the others disagreed. They appointed Allier to deal with German demands for explosives. During one encounter, Allier later recalled, the Germans asked him directly what had become of the heavy water "seized in Norway by a French officer." If Allier was surprised because the Germans had still not figured out that *he* was that officer, he concealed it and told his questioner that the heavy water had been sunk on its way to Great Britain. This notion was accepted by the Germans, who thereafter left Allier alone.

But they did not leave the source of heavy water alone. A delegation of German scientists went to Norsk Hydro and were appalled by the deliberate slowness of its production; the output was about 300 pounds a year, while calculations showed that 5,000 pounds of heavy water would be required for the building of an atomic bomb. The Germans tried to make adjustments that would upgrade the facility and its annual production capacity.

Against the late-August 1940 ordeal of German night bombing, Great Britain's armor consisted of its radar warning system, barrage balloons, searchlights, fewer than a dozen antiaircraft guns guided by gun-laying radar, and a handful of night fighters—ordered the previous summer and just arriving at airfields—augmented by the not-yet-fully-tested "meacons." In the two months since Reg Jones had briefed Churchill, the British had worked at top speed to devise ways to jam and mislead the Knickbein system. Jamming was easier than sending out false signals but could be more readily recognized by the Germans, and the cabinet feared that a "jamming war" could sap the utility of Britain's own defensive radar. Nine Home Chain stations were fitted with equipment to send out masking beacons, and British reconnaissance managed to locate Knickbein-originating stations at Cherbourg and Dieppe, permitting early detection of the German beams and the bombers they guided.

But defense against a bomber by antiaircraft fire was difficult at best. From a height of 20,000 feet a bomb drops to the ground in 35 seconds, and if the bomber is traveling at 440 feet per second, that means it is 2.5 miles away from its target when it releases the bomb; so the defending unit must fire at the bomber when the plane is farther away than that and before it can release the bomb, and may have to "lead" the plane by more than a mile with its fire.

While the task of hitting a bomber at 20,000 feet from the ground was difficult, the British air-defense task was actually assisted by German orders that their bombers attack from between 21,000 and 23,000 feet. According to Adolf Galland, this height slowed down the bombers and made them better targets for the defending planes, since Home Chain radar could detect the attackers at three to four miles away. Had the German bombers flown at a lower altitude, British radar would not have found them early enough to prevent them from reaching their targets.

An equally decisive factor sapping the effectiveness of the German night bombing emerged from a cycle of events that began on August 24, when a German bomber aiming for fuel tanks along the Thames mistakenly bombed London. In response, Churchill ordered four bombing raids on Berlin. Their impact on the German capital was slight, but their impact on Hitler was large. The Berlin raids proved to him that British bombers could get through to Germany, possibly with bombs that contained chemical or biological agents; such potential for retaliation served to deter Germany from using gas in its own bombs. The British raids on Berlin also caused Hitler to redirect the Luftwaffe's planes against London rather than continuing their successful attacks on British airfields and plane-manufacturing facilities. Galland and other aces protested the switch in targets. Sparing the British aerodromes and factories would ultimately prove to be a significant German mistake in the Battle of Britain.

At teatime on September 7, 1940, waves of 300 German bombers escorted by 600 fighters started the British capital burning. Many of the bombs were two-pound magnesium incendiaries that produced fire all out of proportion to their size. A second set included chemicals that detonated when sprayed with water. A third were packed with delayed-explosive devices so that they would kill firefighters and rescue teams.

During the first week of persistent bombing of London, many industrially important facilities were hit, two unexploded bombs lodged in Buckingham Palace, and everyday life was seriously disrupted. September 15 occasioned the most concentrated aerial battles of the war. One failing of British radar just then was that while it could pick up individual German planes as they rose from their bases, the system could be overwhelmed by large masses of planes and fail to properly identify them. This day Göring threw phalanxes of bombers and fighters at Great Britain, confusing the Home Chain. However, the multiple, slow-moving German bomber squadrons also provided easy targets for British Spitfires and Hurricanes that penetrated the German fighter umbrellas. The Spitfire was able to outmaneuver all the German planes but the Me 109. Together, Spitfires and Hurricanes shot down 56 German bombers on September 15.

Because the Luftwaffe had so clearly failed to break the British air defenses, on September 17, 1940, Hitler indefinitely postponed the invasion of Great Britain. But he did not stop the night bombing of London and other British cities; rather, he increased it, to sap what he styled the British "will to resist" and also to conserve the Luftwaffe, which in daylight flights had lost nearly twice as many as aircraft as had the RAF.[3]

Night bombing was directed by Knickbein and by X-Gerät, another system of intersecting beams that had been in development since 1933. Mounted in the planes of Kampfgruppe 100, this more accurate system allowed that group's small bombers to serve as pathfinders, locating a target and dropping incendiaries and flares to guide the main fleet that followed. They set aflame the City of London, the financial district of the capital. The dastardly nature of the German incendiaries and delayed-action bombs firmed the British resolve to retaliate in kind when they bombed Germany and spurred the Allied development of even more intensely incendiary bombs.

British radar labs tried to find ways to counter X-Gerät and to foil new acoustic mines introduced in the fall of 1940. Reg Jones had a

[3]The eventual totals of the Battle of Britain would show that 1,733 German aircraft and 915 British fighters were destroyed.

moment of grim satisfaction on realizing that both new weapons had been predicted by the Oslo Report; then he wrote a strong note to his superiors that the existence of Knickbein proved that the Oslo Report's "source was reliable, and he was manifestly competent." The work to counter X-Gerät was not completed in time to prevent devastating German raids in mid-November, in which 60,000 of Coventry's 75,000 buildings were badly damaged or destroyed.[4]

Foiling the acoustic mine, though, was quickly done: a combination of construction jackhammers and loudspeaker systems, deployed over the sides of ships, did the trick. It was wonderful to see what first-class scientists could do when they set their minds to military-related problems. Cambridge mathematician E. F. Collingwood speculated about what sorts of mines the Germans might introduce in the future and helped establish of a special unit in the Naval Mine Design Department to investigate the possibilities. The Collingwood unit's first guess was that the Germans might make use of water pressure as a triggering device: when a ship passes over a body of water, there is a slight decrease in the pressure directly below—this is what causes a ship to be sucked against the side of a dock during docking. Researchers then discovered that there were all sorts of ocean phenomena whose parameters were not known, such as the sound frequencies of various ships in a harbor and the ability of sea water of various densities and saltiness to transmit sound or pressure or magnetic fields. They began basic research to learn such things so that countermeasures to future mines could be developed, and in the process they expanded the store of general knowledge about the properties of the sea and how objects acted in it.

The average time lag between the moment that a scientist or tech-

[4]A report persisted for decades thereafter that the British had known about the Coventry raid in advance, through Enigma decrypts, but had not warned the city or sent aircraft to defend it, in order not to compromise the code break. Recent scholarship has shown that this was not the case.

nologist was put onto a problem and the production of a solution was about six months. By early winter 1940, British science-based research efforts hurriedly begun in May and June were starting to produce results—for instance, from the mind of Edward Terrell, a thirty-eight-year-old barrister and inventor of a revolver clip for rapid loading, who had been called up by the Admiralty in June to take a position similar to Jones's, looking into what the Germans were doing in science adapted to war. After ordering a uniform from his Savile Row tailor, Terrell went to work. He was aghast to discover that sailors in gun turrets and command posts were being killed at an alarming rate by the Luftwaffe's machine guns, because the steel armor plate did not adequately stop bullets; in an earlier court case Terrell had learned of granite's high crushing strength, and he used this as the basis for reinforced-concrete armor that better protected crews from aerial gunfire. To prevent the enemy learning its composition, it was called "plastic armor," though it contained no plastic. Next Terrell tried to learn why pilots of German dive-bombers did not black out at the bottom of their dives, as British pilots did; a captured Junkers 87 revealed an ingenious automatic pull-out button that brought the plane out of a dive even if the pilot temporarily lost consciousness. It was relatively easy to reproduce the device for RAF planes but more difficult to convince Royal Navy ships to take evasive action when attacked by dive-bombers.

Terrell, Jones, and other scientifically trained members of the military establishment were asked to evaluate innumerable schemes for war-related inventions sent in by the general public. Most were crackpot and immediately dismissible. Only one in twenty got as far as a test: one of those tested was a mirror in the nose of an aircraft to simulate the flashes of gunfire and blind the gunners of the opposing craft; it never accomplished that task. More promising were devices conceived by military researchers to solve problems that civilians never contemplated, such as eliminating smoke from the stacks of cruising ships so that German spotters with binoculars could not see the plume to locate a freighter from as far away as twenty miles. Concealment was achieved by forcing extra oxygen into furnaces so that fewer unburned particles were disgorged up the stacks.

The most important scientific contribution that began in Great Britain that fall was Operational Research (OR), loosely defined as the application of the scientific method—by scientists—to solving the problems of conducting a war. While not as exciting or instantly understandable as a V-2 rocket or a proximity fuze, Operational Research would eventually have a greater impact on the war. One of its early and best practitioners, P. M. S. Blackett, was soon pointing out that the brass mistakenly envisioned new war-related devices as "springing like Aphrodite from the Ministry of Aircraft Production in full production, complete with spares, and attended by a chorus of trained crews. . . . [But] relatively too much scientific effort has been expended hitherto in the production of new devices and too little in the proper use of what we have got." OR was first named and advocated by Watson-Watt, in an attempt to evaluate the efficacy of radar. From radar the technique and mind-set of OR broadened out to the task of assessing the worthiness of all sorts of weapons and supply systems. A manifesto for OR appeared in an anonymously written book in late 1940, entitled *Science in War*, actually penned by Blackett, J. D. Bernal, Solly Zuckerman, and C. H. Waddington, all early pioneers. Their basic contention:

> The use of [technologically advanced] weapons and the organisation of the men who handle them are at least as much scientific problems as is their production. . . . Failure to understand the factors contributing to victory or defeat and the degree to which each contributes, removes any secure ground for organising further success.

From a first level of assessing the efficiency of an existing system, OR advanced to testing the best way to use a new weapon or system prior to its general deployment, and eventually to writing the specifications for new systems and for proper coordination among the groups making the weapons and those operating them.

Bernal was attached to a new department of the Ministry of Home Security, near Oxford, and was working on the physics of explosions and the ability of various buildings and structures to resist shocks. His enthusiasm for determining how to protect people from the bombs convinced Zuckerman to become involved in studying the risk of concussions to those in underground shelters, and then to an overall study of bombing and its effects that would later be translated into a cornerstone of Allied military bombardment policy.

As the politicians and the military were beginning to understand, everything about modern warfare was accelerating—from the speed of the aircraft to the spin of the projectiles, from the rapidity of production to the accumulation of information—and mathematics was the key to controlling those factors. And contemporary scientists, whether theorists or experimentalists, had mathematics in their blood, because advanced mathematics had become integral to every modern science, from cell biology and polymer chemistry to nuclear physics and the study of radio waves. To operate at a high level of complexity in warfare, mathematics was an indispensable tool, which meant the military had to accept a partnership with science—an increasingly equal one.

As the bombs of "the Blitz" fell nightly on Great Britain during the fall of 1940, and as supply ships heading for Great Britain were regularly being sunk by U-boats—the toll reached 250,000 tons a month—President Roosevelt signed a secret agreement for the United States to fully equip and maintain ten British divisions and to send half of what America's munitions factories produced to the British Isles. Eventually this arrangement would be ratified by Congress as the Lend-Lease program. Roosevelt felt able to rush supplies to Great Britain right away because he had become confident that he would be elected to a third term as president in November.

That month the Radiation Laboratory at MIT was commissioned. It was referred to as such to make outsiders think its mission was not military, since nuclear radiation was not then generally considered a militarily related matter. The Rad Lab had three objectives to develop: airborne interception radar, gun-laying radar for antiaircraft fire, and a navigational aid system. The presence of Taffy Bowen as an expert (and a Briton) was critical, because, as he wrote to Tizard, "The urgency of the situation has not yet been realized . . . and progress is about half as fast as in England." The Rad Lab's pace picked up as Loomis, Lawrence, Bush, and other heavyweights persuaded some of the country's best scientists to join. At an early meeting, Edward Condon asked the group if they knew what a cavity magnetron was; the disassembled parts lay on the table in front of them. "It's simple," said nuclear physicist I. I. Rabi, one of the older men. "It's just a kind of whistle." "Okay, Rabi," Condon replied, "how does a whistle work?" Rabi could not immediately answer the question, but it spurred an interested investigation of the subject.

Rad Lab denizens were mostly young physicists and engineers, many of whom would go on to distinguished careers. A disproportionate number of them were nuclear physicists. Trying to explain why nuclear physicists were able to dive right into radar work and make a difference, Luis Alvarez, a nuclear physicist who as a young man was one of the early Rad Lab employees, later wrote that

> nuclear physicists had greater aptitude for pulsed radar development because we made our living with electronic pulses from ionization chambers and Geiger counters. Radio and electronic engineers, who were more familiar with the quasi-sinusoidal signals used in audio or radio-frequency engineering, abhorred pulses; such signals usually indicated trouble—lightning flashes or sparking.

For the young scientists, being at the Rad Lab was a time out of time. Suspending their previous routines and individual research, they addressed communal tasks of societal importance. Behind windows painted black to discourage onlookers and on a rooftop with a view of the Boston skyline, they became buddies, wrote limericks, and worked until exhausted; the pace of their work and its importance to the war invigorated them. Their schedule was ambitious: a prototype microwave system by early January 1941, an installation in a B-18 by February 1 and in a night fighter before March 1.

Not all the American efforts were as well organized or well funded. To begin work on the "influence fuze," later known as the proximity fuze, Merle Tuve of the Carnegie Institution had much more limited resources. He had to buy black powder commercially from a Georgetown shop, and he and his compatriots at the Carnegie lab had to make their own vacuum tubes, then take their contraptions out to a friend's farm in rural Virginia for test-firing. Tuve and his colleagues felt unqualified to do war work, but did it anyway. Often, their naiveté showed through. Tuve asked Raytheon for $10,000 worth of hearing-aid vacuum tubes, only to receive a lecture from a Raytheon executive that the order was too small to be honored; usually, the executive told Tuve, they used the first batch of $100,000 worth of tubes just to get the bugs out.

Tuve was also a member of the Briggs Committee, the American group concerned with nuclear research that might lead to making an atomic bomb.

The Tizard Mission had conveyed to the Americans and Canadians a sense of what was being done by the MAUD Committee, but there was no real exchange of information among the several countries' nuclear researchers. Some teams in the West, and in Germany, were making progress on the next obvious step in bomb construction: determining whether another isotope could substitute for U-235, since so many difficulties were inherent in separating

U-235 from its ore. Shortly this search would lead to Glenn Seaborg's creation of plutonium and to a relatively rapid route to producing a self-sustaining chain reaction. In the general public there was a frisson of fear about the possibility that the Nazis might develop an "atom bomb," introduced in such pieces as an article in the *Saturday Evening Post* of September 7, 1940, entitled "The Atom Bomb Gives Up."

In France the Germans imprisoned Joliot's mentor, Paul Langevin, which occasioned protests from scientists in a dozen countries. On November 11, the anniversary of the armistice that ended the Great War, five thousand Parisian students took to the streets to protest Langevin's detention, many carrying two poles—in French, *deux gaules,* a reference to Charles de Gaulle, leader of the French in exile. Foolishly, Joliot climbed a gate to place some flowers around the Sorbonne's monument to the war dead; a caretaker recognized him, and Joliot was reprimanded, not only by the administrator of the Collège de France but also by several colleagues on the science faculty, including Langevin's son-in-law, Jacques Solomon, who did not want Joliot to expose himself to undue scrutiny just when they were beginning to organize for true resistance. Joliot's refusal to open his laboratory until Langevin was freed reduced his incarceration to house arrest. Kapitsa offered Langevin a position in the USSR, but Langevin, too, felt he must not leave France.

Occasionally Bothe would come to Paris to work at Joliot's cyclotron. But as soon as Bothe began an experiment, the French chief mechanic would make an excuse to go to his workshop and on his way would shut off a water tap in the cooling system; within minutes the experiment would fail. This pattern was repeated many times, but sabotage was never suspected. Science historian Spencer R. Weart, who studied Joliot's work, concludes that what did *not* happen in the Joliot lab hurt the German nuclear program:

> Such little failures, together with poor progress in laboratories in Germany, had a large significance. While the Germans suspected that plutonium existed and that it might possibly be used for a compact and overpowering bomb, they never got any idea of the element's properties nor even any proof of its existence. They failed to push this line of investigation, which was the quickest way to a bomb.

Adding to the "little failures" was a mistake in measurement made by Walter Bothe, which rejected the use of carbon as a moderator for the atomic pile, in favor of heavy water. This mistake was compounded by Heisenberg's unchallenging acceptance of heavy water and his rejection of graphite; the latter could have brought Germany's nuclear program to the brink of making an atomic bomb.

German work toward a bomb was slightly hampered by the the Szilard-led effort to halt publication of American and British results, though the effect was like shutting the barn after the horse had fled, since the 1939 Hahn-Strassmann and Meitner-Frisch articles had spurred the publication of a hundred further articles on aspects of nuclear fission. Even without knowing the latest experiments of Fermi and Szilard, Heisenberg and von Weizsäcker duplicated the American results, and, on another aspect of the problem, Paul Harteck came to the same conclusions as Frisch.

In Japan a survey by the Army verified that enough uranium could be obtained from deposits in Japan, Korea, and Burma to make a bomb, and in October 1940 the Riken Physical Chemical Research Institute appointed Yoshio Nishina head of an effort to build one. The 28-ton-magnet cyclotron completed for Bohr's 1937 visit was now a generation out of date, so Nishina and scores of younger physicists worked to finish a 210-ton-magnet cyclotron at the Riken Institute from plans donated by Ernest Lawrence; in December 1940 the Riken cyclotron was ready to begin measurement and bombardment work.

By then Japan had signed a tripartite agreement obligating Germany, Italy, and Japan all to go to war against any other country, including the United States, should that country attack any of the three. Perhaps in response to the possibility, the Japanese Navy replaced its communications code in December 1940.

American cryptographers had recently finished cracking the previous code, and now they had to start all over again; rather than assign the new task to cryptographers in Washington, however, the brass decided that the smaller group on Corregidor should work on it while Washington concentrated on the Japanese diplomatic cipher. Japanese codes were fundamentally different from those made on Enigma machines; they came from the permutations of the "wipers," switches found in all contemporary telephone systems; each wiper unit was six levels wide and twenty-five steps long, providing a large number of possibilities for substitution; in addition, the Japanese language provided another layer of difficulty for Americans trying to crack codes made by the "Purple" machine.

When Chinese counterattacks killed 20,000 Japanese soldiers in China during the latter part of 1940, the Japanese military began to feel they must widen the war in order to win it. In terms of scientific preparations for helping to win the war, despite the recommendations of the 1938 report of the Japanese Science Council, in late 1940 there was still a lack of coordination for this purpose among government ministries and among the departments of the armed services. A "General Plan for the Establishment of a New Science and Technology System" circulated in the Japanese cabinet, but there was as yet no general agreement on what war-related work it should cover nor who would lead it. A year before the attack on Pearl Harbor, Japanese scientists still had almost no part to play in their country's military preparations for the wider war.

| CHAPTER SIX |

Toward Pearl Harbor

On December 7, 1940, James B. Conant presided over a symposium held at Harvard on the relationship between the university and defense work. For Americans the shooting war in Europe was far away, and the Atlantic and Pacific Oceans were still considered adequate barriers to protect the United States from harm. Conant, an advocate of greater U.S. involvement, was pleased to note that alumni and faculty had begun to swing from adamant isolationism to moderate participation, now willing, for instance, to give academic credit for training in aviation and to establish a Red Cross/Harvard hospital in Great Britain.

In early 1941 Conant rather than Bush was selected to head a delegation of scientists to Great Britain, the counterpart of the Tizard Mission of the previous summer. Prior to leaving, Conant testified before Congress on behalf of President Roosevelt's Lend-Lease bill, as did New York City's mayor, Fiorello La Guardia, and Wendell

Willkie, the 1940 Republican candidate for the presidency; opposition to Lend-Lease was led by prominent isolationists Charles Lindbergh and Henry Ford. Conant's testimony in favor of Lend-Lease stood him in good stead in London. In March 1941 he witnessed the terrors of the Blitz firsthand—luncheon with Churchill in the bomb-proof basement at 10 Downing Street during an air raid was just a start. During another raid he had dinner at the private estate of Lord Stamp, a noted economist and chairman of the railway board, who insisted on remaining in his own home during the bombing rather than go to a shelter. Two months later Stamp and his family were wiped out by a direct hit on the estate; the event so enraged a surviving son, a biologist, that the next Lord Stamp left a safe job and volunteered to work at Porton Downs in the dangerous field of chemical- and biological-warfare agents.

At Conant's private lunch with Lindemann at his London club, Conant learned of the growing British interest in nuclear fission. When Conant "wondered if it was wise to devote the precious time of scientists, with the German threat so critical, to a project which could not affect the outcome of the war," Lindemann told him that bringing together a critical mass of refined uranium could produce "a bomb of enormous power." It was the first that Conant had heard about "even the remote possibility of a bomb," and he decided not to make further inquiries, leaving that for higher-ups to do.

Conant was present in Great Britain when word came in from the Mediterranean that radar had been used successfully in a big naval attack: British planes and ships had located by radar, and then sank, three Italian cruisers and several destroyers, practically wiping out the Italian Navy. Germany would use the evidence of this attack to verify the existence of British seagoing and airborne radar; after it, Germany accelerated the transfer of its own radar, and what information it possessed about British radar, to its Japanese allies. Around this time as well, Japanese spies in New York reported the details of the British radar system mounted aboard the French liner *Normandie,* then docked in New York's harbor.

Upon Conant's return to the United States, President Roosevelt received him at the White House. The Harvard leader told his university's most prominent graduate of the need to send Americans to Great Britain to learn about radar; "to my astonishment," as Conant later wrote, he found that "the President was almost totally ignorant of the functioning of radar," perhaps because published accounts of the Battle of Britain had given "hardly a hint" about the use of electronic devices. In response to a direct presidential request, Conant briefed Roosevelt on radar, after which "the President went to his Cabinet meeting full of enthusiasm for my idea."

Conant had to make amends to Stimson and Bush for having gone out of channels, but they forgave him. He was shortly appointed chairman of the NDRC, when Bush became director of the Office of Scientific Research and Development (OSRD), which was formed to supervise the NDRC and a Committee on Medical Research.

The birth of the OSRD overlapped the final stages of a classic and inevitable challenge to Vannevar Bush and civilian involvement in military research. The challenger was Vice Admiral Harold G. Bowen, who as director of the Naval Research Laboratory was a member of the NDRC's governing council. A man who by his own admission was very tough, liked to speak his mind, and had made influential enemies in the Navy, Admiral Bowen had wanted all research put under his command in a new "Naval Research Center." The emergence of the NDRC in 1940 had eclipsed that notion, and Admiral Bowen was further angered when the NDRC chose to concentrate on the cavity magnetron brought by the British rather than on the NRL's previously developed longer-wave radar. The last straw for the admiral was the matter of who would control research on antisubmarine warfare. The Secretary of the Navy asked the National

Academy of Sciences (NAS) to evaluate the Navy's antisubmarine research and to advise on potential new methods for detecting submarines. That report, which advocated a national commitment to speed up the research, with civilians intimately involved, reached Bowen in January 1941. In a note to the new Secretary of the Navy, Frank Knox, Bowen rejected the report's advice; he contended instead that civilian scientists should work only under the direct supervision of the military, and he intemperately charged that the only reason for Knox to adopt the NAS recommendations "would be on account of the pressures exerted by certain well-known scientists." Perhaps Bowen believed that resisting such pressure would appeal to Knox, a former Rough Rider under Teddy Roosevelt.

Bush could not simply ignore Bowen's report to Knox, which denigrated civilian contributions to military research; he had to squash it. He first successfully appealed to Bowen's rivals in the Navy, the leaders of the Bureau of Ships, to request NDRC participation in antisubmarine-warfare research, then asked Knox to have Jerome Hunsaker of MIT, a former naval officer and the treasurer of the NAS, review the Navy's antisubmarine research and Bowen's handling of the NAS report. Hunsaker's critique dismissed Bowen's plan for a Naval Research Center and recommended instead establishing a new office to coordinate the Navy's work with civilian scientists. Knox agreed. He appointed Hunsaker the new coordinator and filed an unsatisfactory fitness report on Bowen that effectively blocked the admiral from further advancement.

Bush won out over Bowen in 1941, and by creating the OSRD he further diminished the military's autonomy in the making of weapons, since the OSRD charter permitted this civilian-controlled agency to develop and test a weapon even if the military did not approve. Unlike the NDRC, which had drawn money from the president's emergency fund, the OSRD enjoyed direct funding from Congress and was on firmer legal ground. As Bush had hoped, the immediate result of its creation was to push the military services into greater cooperation with civilian researchers.

In 1940 and early 1941 Canada was a belligerent, pledged to support the mother country, to whose aid it was sending troops, ships, planes, and other war matériel, but mobilization was far from complete. In terms of scientific research related to military matters, there were in Canada two sets of patrons: the government and a small group of private businessmen, including the distillery magnate Samuel Bronfman, who contributed substantial funds for work that the government did not yet see fit to underwrite.

Sir Frederick Banting, codiscoverer of insulin, used some of these latter funds for work on the toxicology of poison gases, the use of pressure suits to prevent pilot blackouts, and the beginnings of biological- and germ-warfare projects. After a meeting with Tizard—who elated the Canadian scientific community by visiting Ottawa before going to Washington, D.C.—Banting read everything he could find on radioactivity, a subject about which he had previously been ignorant, and within a week was jotting down ideas about the relationship between "misplaced energy emanating from radioactive organic elements" and cancerous cells, and urging colleagues to back uranium experiments, which they did. It was evidence that Banting's influence and enthusiasm served to galvanize all sorts of Canadian military research.

In October 1940 Banting had reported to the War Technical and Scientific Development Committee that the bacterial warfare experiments had "reached the point where field work must be carried out on the means of distribution from the air." Shortly he sprinkled a stream of sawdust from an aircraft to measure how particles would spread if they were infected with bacteria. Banting received a "green light" to go into production of war bacteria. Their use was contemplated only as a retaliatory measure, but the stock had to be field-tested and manufactured in sufficient quantities to be ready if needed. As the year's end approached, Banting worked long hours, drank more, and expressed the wish to take direct part in the fray. In early February 1941 he left for Great Britain to check on scientific developments there, including a CBW facility newly begun at Porton Downs, hopping a ride on one of the bombers being ferried across the Atlantic. The plane crashed, killing all on board.

By January 1941 the German euthanasia program had been used to kill about 50,000 mentally and physically "defective" people, many of them children. One result of the growing dimensions of the program was that it became public knowledge. Hitler was jeered at a train stop by those objecting to the euthanasia program, and he ordered it halted. It was soon restarted, with greater secrecy.

Some doctors associated with universities located near concentration camps asked to have a few prisoners already condemned to execution as subjects for medical experiments designed to help German pilots survive extreme conditions. The subjects made available were not murderers but Jews and other inmates; they were exposed to freezing mixtures, high heat, vacuums, high pressure, and other tortures until they expired, their death agonies captured on film as a record of the research.

As plans for the invasion of Russia firmed, the SS revived the Special Task Forces that had killed tens of thousands of Jews in Poland; their new assignment would be to kill Jews and Communists in Russia. Chemists at a company in which I.G. Farben held a controlling stake were tinkering with the gas Zyklon-B, a prussic acid used in pesticides, at the request of the SS. The pesticide version had an odor-causing component to warn people from the area being sprayed; the chemists were told by the SS to remove that component so the gas would be odorless as well as deadly.

The Rad Lab at MIT missed its January 1941 deadline for a prototype microwave radar system, but on February 7 a group of physicists and engineers, mostly new recruits from Berkeley like Luis Alvarez, set a prototype on the roof of an MIT building and managed to detect a passing aircraft and follow it by radar for more than two miles. Seven weeks later a B-18 bomber—and a nervous group of scientists aboard, including E. G. Bowen—took a flight during which their prototype air-to-air radar detected a target National Guard airplane at a distance of between two and three miles, a ship on Cape Cod at nine miles, and then, as the scientists pushed their luck, several submarines in the yards at New London, Connecticut. Similar results had recently been obtained by the British, but with longer-wave sets that were very cumbersome and provided depictions of target objects that were not as sharp.

The Rad Lab's two other projects, a long-range navigation system—eventually to be called LORAN—and a new antiaircraft gun-laying radar system, took longer. Both were based on Alfred Loomis's ideas. While Loomis was taking a shower, a vision came to him of a grid in the sky, composed of synchronized, crisscrossing radio pulses that could very accurately direct an airplane. The idea of conical scanning to keep a target always in focus, the heart of a new system for control of antiaircraft fire, was also Loomis's but had occurred also to scientists in Great Britain and Germany. The complementary notion of taking the control of antiaircraft fire out of human hands and letting it be done by computing circuitry was developed by Bell Labs and by other Rad Lab scientists.

Also at MIT was Harold Edgerton, the world's leading expert on high-speed photography. The Air Force asked Edgerton and his associates to design a camera that could take photos of enemy installations at night. In May 1941 Edgerton installed a huge camera aboard one plane, a flash-making device in another, and arranged for them to fly in tandem at 2,000 feet in the air, during a very dark night, and take pictures of familiar objects—an airfield in Boston, Yankee Stadium in New York, and General Hap Arnold's home in Washington, D.C. Arnold seemed unimpressed, because the camera could not operate at the 25,000-to-30,000-foot altitude that would be above the flak of the antiaircraft guns Germany was using to deter British bomber raids against their homeland. The night-flash machinery was put on hold for the next two years.

On May 15, 1941, Henry Tizard and officials from the Ministry of Aircraft Production and the RAF witnessed the first flight of a jet-powered aircraft, with a Whittle turbine—a project Tizard had backed for several years, despite skepticism from many scientists, technicians, and politicians. Tizard viewed the jet aircraft as one of the few technical developments that could tip the balance in the war. A thousand of the aircraft were envisioned, but it was uncertain when enough prototypes would be available so that extensive testing could be done. Had Tizard still been highly influential in the government's research and development decisions, he might have pushed for more rapid development and production of jet planes. But those days had passed, and the men at the top in Great Britain continued to believe that the war was being won by conventional propeller planes.

The German invasion of Soviet Russia began on June 22, 1941, precisely one year after the armistice in France. This undertaking was a fulfillment of Hitler's enmity to Communism, a drive for German lebensraum, and an opportunistic attempt at a quick conquest while the Luftwaffe was being rebuilt for a second run at Great Britain. To his generals Hitler characterized the fight against the Soviet Union as "a war of extermination." It would become just that for Germany as well as for the Soviet Union.

No other aspect of World War II, not even the dropping of atomic bombs on Hiroshima and Nagasaki, produced nearly as much devastation as this terrible conflict of the Soviet and German dictatorships. Of the 20 million killed on the Soviet side, most were noncombatants, starved to death, or shot or gassed by the SS, whose orders were to execute without trial local Soviet political leaders and known Communists. In fourteen days the SS killed 10,000 Jews at Kishniev, herding the victims into pits and machine-gunning them. Impressed by such efficiency, in midsummer of 1941 Göring instructed the leader of the Reich Security Service, on Hitler's behalf, "to make all necessary preparations as regards organization and actual concrete preparations for a general solution of the Jewish problems within the German sphere of influence." This order accelerated the building of gas chambers and the rounding up of people to be executed in them.

More than a million Red Army soldiers were taken prisoner; most of them, too, were killed after capture, to avoid the Germans' having to feed them. German troops were also ordered to burn excess Soviet stocks of food, which contributed to the millions of deaths on both sides.

Hitler and his military chiefs underestimated by far the adverse effect on their forces of the immense distances to be traversed in the USSR. This, even more than "General Winter," was the principal cause of the German defeat. While Hitler's armies were mobile, they were not truly mechanized; only the tanks had tracks that permitted them to travel in areas where there were no roads or where "General Mud" hopelessly mired German supply vehicles and the tens of thousands of horses pulling them. The prior inability of the Third Reich to direct the talents of its scientific and technical institutions to the tasks of mechanizing its armed forces significantly contributed to the military failure. They failed to accomplish such obvious tasks as perfecting antifreeze for tank radiators, or of heating apparatuses to keep gasoline liquid, or of liquid solutions to prevent synthetic rubber from losing elasticity in the cold.

The great distances of the Russian plains and the central control of Soviet society contributed to the Soviet Union's completion of one of the great technical feats of the war, the large-scale transfer of crop production, industrial facilities, and scientific institutes from west of the Ural Mountains to east of that natural barrier. Not all the scientific laboratories survived the move. The uranium laboratory, for instance, was shut down and its nuclear physicists scattered to work on non-nuclear projects. Some 500 factories were shifted between 1939 and 1941, while the remainder were moved after the German invasion had begun. Still, during that invasion the USSR lost a third of its industrial facilities and half of its agricultural lands.

Russian scientists had a hand in helping their people survive: in Leningrad during the long siege they concocted a bread containing 10 percent cellulose, 10 percent cattle feed, and 4 percent paper and flour dust mixed with the regular flour. Tractors were modified to run on fuel derived from wood, peat, and straw briquets, and peat was used as a substitute for commercial, nitrate-based fertilizers. In general, however, the Soviet regime's earlier purges of its scientific and technical talent hurt its efforts to survive the German invasion and to win the military clash with the Nazis. This was when the first installment of the bill came due for Lysenko's tinkering with the harvests: widespread shortages of grain, due in part to improper planting techniques in the areas of the Soviet Union not overrun by the Germans.[1] Soviet radar did not work very well, Soviet medicine made few advances in restoring the wounded to battlefield readiness, and Soviet technical institutes were very slow in translating laboratory breakthroughs into reproducible machinery. Among the few technological successes were antitank mines; made of nonmetallic parts and placed inside wooden boxes, they could not be

[1]Despite the shortages for which he bore some responsibility, Lysenko was soon awarded a cash prize for distinguished work. Other prizes went to Kapitsa, to astronomer Sergeo Orloff, and to the designers of Soviet tanks and planes. Lysenko's best project was a 700,000-acre potato patch.

detected by magnetic means and were effective in countering German tanks.

As the war had widened, the response in Great Britain—and even in the United States, which had not yet become a belligerent—was for the leaders of civilian scientists to increasingly assume positions of responsibility on government-related projects and for individual scientists to agree to have their talents and resources appropriately directed to various war efforts. But in Germany the leadership of science had become thoroughly Nazified, and when the crisis of the war in Russia occurred, that leadership took the opportunity for self-aggrandizement. An alliance between Alfred Rosenberg's Ministry for Occupied Eastern Territories and the Kaiser Wilhelm Gesellschaft (KWG), begun earlier for the exploitation of Poland's scientific resources, now expanded. German scientists were sent to oversee—and in some instances to loot—Russian efforts that they had long envied, such as the work of the plant-breeding research stations formerly led by N. I. Vavilov, then slowly dying in a Soviet prison. The collaboration between Rosenberg and the KWG also founded KWIs in Bulgaria and Greece. In exchange, leading KWI scientists received military deferments—at a moment when more than 40 percent of KWI staffs had already been drafted.

On the thirtieth anniversary of the founding of the KWG, in mid-1941, the Nazi Party's newspaper featured articles touting the scientific community's cooperation with the war effort. It highlighted the practical KWIs' performance of tasks related to the production of arms, the enrichment of poor iron ores, the development of new alloys of steel and light metals, the creation of new processes for transforming low-grade coal into fuel oil, and the medical enhancement of the recovery of wounded soldiers. Three featured photographs were of the physiochemist Otto Hahn, the practical physicist Ludwig Prandtl, and the eugenicist Eugen Fischer—a trio that reflected how German science had evolved in the Nazi era: some basic research to show the rest of the world that Germans were not barbarians, some very practical research in aviation, and an equal dollop of pseudoscience. The KWG's new president, Albert Vögler, would shortly define the goal of KWI research as "the quickest interpretation of results for the war."

Germany's military chiefs had envisioned a four-to-six-week campaign in the Soviet Union. As it lengthened into a costly attempt to hold enormous stretches of territory and the chiefs were forced to focus all their attention on the eastern front, at home all sorts of potential weaponry and applied-science research languished. The EZ/42, an airborne gunfire director capable of improving a fighter pilot's ability to shoot opposing planes out of the sky, was not given adequate field trials. Several forms of radar, though developed in the lab, fell behind the British efforts to counter them. German jet-engine development slowed at just the point when, with new funding and added resources, it would have begun to provide a potential war-winning weapon. Designs for new submarines that would have dominated the seas were shelved for lack of high-level interest.

During the summer of 1941, Senator Harry S. Truman expressed a point of view held by many in the United States: elation at the spectacle of the Russians and Germans killing each other and a belief that the United States should urge on whichever belligerent was lagging behind. Interventionist fever, which had risen during the winter and spring in the United States because of the obvious danger to Great Britain, fell after Germany invaded Russia, an action seen as taking the military pressure off Great Britain for the immediate future. Isolationist sentiment was so strong that the renewal of the draft, the Selective Service Act, passed the House of Representatives only by a one-vote margin, 203 to 202.

The American military, however, understood that full American involvement in a shooting war was almost inevitable and in small and large ways continued to prepare for it. For years one of the military's chief civilian consultants had been Charles Stark Draper of MIT. Draper was a man of many engineering trades; he held the distinction at MIT of having taken the most courses in the most departments without receiving a doctorate—and, some professors said, of knowing more than they did—until his Ph.D. was finally awarded in 1938. He was also a skilled stunt pilot who contributed many ideas and devices to maintain stability in aircraft, particularly a gyroscopic rate-of-turn indicator. Early in 1941 he was asked to apply that indicator and his own ingenuity to antiaircraft guns, a task that Draper characterized as having to provide "the solution of a problem which at first glance seems too complicated to solve," hitting a target that was moving faster than the hand-eye coordination of a human being could follow.

During the summer of 1941, on a firing range, Draper demonstrated for the Navy a gunsight for ship-based antiaircraft fire. To be called the Mark 14, it used a rate gyro as a computer to rotate a mirror that allowed the gunner to continuously track the target and properly lead it with fire—while correcting automatically for the effect of gravity on the bullet's trajectory, for wind speed, and for the "velocity jump" due to the angle at which the gun was being fired. A competition was held at the Dahlgren firing range between experienced gunners operating prior tracing equipment and an untrained seaman whose fast reflexes had stood him in good stead as a boxer, operating with the Mark 14. Shooting at a towed target on a steady course, the competitors were even; with a rapidly moving target, the boxer and his Mark 14 did much better than the experienced crew. Thousands of the Mark 14's were immediately ordered.

By summer 1941 the Battle of Britain had been won by the RAF and the Home Chain Stations. British laboratories had solved the puzzles of Knickbein, X-Gerät, and of the recently introduced and more ingenious Y-Gerät. British factories, spurred by Beaverbrook, had turned out enough fighter planes to supplement those sent from North America, so that never again would the Luftwaffe seriously pierce the air curtain over Great Britain.

Reaching for a way to describe the work of radar in the Battle of Britain without giving away the top-secret details of the technology, Lord Beaverbrook cited in a radio address a story by Pushkin that had been used as the basis of a Rimsky-Korskov opera, "Le Coq d'Or." There was a land threatened by an enemy but its king did not know what direction the enemy would come from. An astrologer gave him a marvel, a "magic bird" of gold. "Set it upon the highest spire," the astrologer said, "and it will keep a faithful watch upon the farthest borders of the realm. When no danger threatens, it will remain quiet. But at the first approach of enemies, my Golden Cockerel will spy out their presence, he will lift his comb, flap his wings, turn his beak in the direction of the raiders, and crow 'Cock a doodle-doo—Beware!' "

That summer British Intelligence succeeded in deciphering the German naval code: Ultra decrypts helped the British move their convoys around in ways that lowered the average tonnage lost each month from 280,000 to 120,000. The decrease in tonnage lost to the U-boats, however, was a chimera that concealed the growing strength of the U-boat fleet. As would later become apparent, the decline in sinkings was not primarily due to Ultra but to the greater distances that the U-boats had to patrol and to a Hitler directive that the U-boats avoid targeting American ships so that the United States would not have an excuse to enter the war. Throughout 1941 German repair docks continued to fit out U-boats with new equipment, torpedoes for attacking escort ships in a convoy and improved deck guns.

In Great Britain several strands of nuclear research were brought together in the summer of 1941. In Cambridge, Joliot's former associates, working with their British and German-émigré counterparts, showed that a nuclear chain reaction could be achieved; at two other universities, Frisch, Peierls, and Chadwick confirmed that such a reaction, if made into a bomb, would have enormous explosive power; and at the commercial firm ICI, progress was made on the chemical-engineering problems associated with isotope separation and other aspects of bomb production. In July 1941 the MAUD committee—with the sole dissent of P. M. S. Blackett—submitted a report that an atomic bomb could be produced in two years, at a cost of £5 million.

This "it can be done" report found its way to Henry Tizard, not because he had any official position in regard to the committee—he no longer did—but because he was known to be hardheaded and practical. Tizard agreed with the recommendation to go ahead but pronounced himself still a "skeptic" because the physics had not been proved, because the construction would require a great deal of money that might be better spent on other projects, and because success in the time frame suggested was doubtful. Lindemann saw the report and came to the same conclusions. On August 27, 1941, he wrote to Churchill that the odds of completing the bomb project in two years were not very good but added, "I am quite clear that we must go forward. It would be unforgivable if we let the Germans develop a process ahead of us by means of which they could defeat us in war or reverse the verdict after they had been defeated." Churchill agreed to have the bomb built, by a project henceforth called Tube Alloys.

At around this time the German bomb project also took a turn, thanks to Fritz Houtermans. A Göttingen physics graduate, in 1933 he had emigrated to the USSR because he was half Jewish and a Communist; imprisoned and mistreated in the Soviet Union, he was returned to Germany in an exchange of prisoners after the Nazi-Soviet Pact of 1939 and had gone to work in the private nuclear-research laboratory of the Baron Manfred von Ardenne. In August 1941 he wrote a long memo on chain reactions, on element 94—the plutonium discovered by Glenn Seaborg—and on isotope separation in the potential manufacture of an atomic bomb. But the other nuclear physicists in Germany considered suspect anything to come out of the von Ardenne lab, and thereafter the German atomic-bomb program veered away from isotope separation, the key to plutonium production, and toward a dependence on natural uranium to produce, and heavy water to mediate and control, the chain reaction. Their mistakes seriously set back the timetable for making a German atomic bomb.

In September 1941 Werner Heisenberg traveled to Denmark and met with his former teacher and mentor, Niels Bohr. For two days the men talked nuclear physics. Bohr received the impression that Heisenberg was trying to learn details of the Allies' progress on nuclear fission and that Heisenberg was directly involved in research on fashioning an explosive nuclear device. He also believed that Heisenberg was trying to convey that Germany had no chance of building a bomb in the near future, perhaps in the hope that this information would dissuade the Allies from going ahead with their bomb project. Heisenberg later denied such intents and stressed that what he had raised with Bohr was the possibility of nuclear physicists on both sides joining together to refuse to build atomic weapons.

While Bohr and Heisenberg were meeting in Copenhagen, British physicist Mark Oliphant was in the United States trying to find out why American nuclear physicists were ignoring the MAUD Committee reports. The answer astonished Oliphant: Lyman Briggs, the chairman of the American Uranium Committee, had hidden the MAUD reports in his safe and had refused to discuss them with Bush, Conant, and the other members. In effect, Briggs had kept the United States in the dark about the possibilities of making an atomic bomb, at just the moment when British research had advanced to the point that even Lindemann had become convinced that a bomb was feasible.

Oliphant visited Lawrence, Bush, and Conant and convinced the most influential American scientists to forgo further work on nuclear power generation and to advocate the use of America's money and resources to build a bomb. A new MAUD report, received by Bush directly on October 3, provided the last bit of ammunition with which he was able to convince President Roosevelt and Vice President Henry Wallace, on October 9, 1941, to accelerate the nuclear-research program and place supervision of it under a committee composed of Wallace, Secretary of War Stimson, Army Chief of Staff Marshall, Conant, and himself. At the urging of Bush, the president also sent a letter to Churchill proposing a joint American and British research effort toward an atomic bomb. Later in the year Churchill rejected the proposal, thinking that British nuclear science was far ahead of its American counterpart.

On October 16, 1941, there was a crisis in Japan, caused largely by the actions of the United States. An alliance had been growing throughout the year among the United States, Great Britain, and the Soviet Union; this new alliance and the effects of the lengthening Russo-German conflict had put pressure on Japan. That pressure tightened after July, when Japan had extended its control over the formerly French Indochina, which brought Japanese forces near to the British stronghold at Singapore, and in response Roosevelt froze Japanese assets in the United States and embargoed shipments of oil to Japan. Great Britain and the Dutch government-in-exile were forced to follow the American lead, and the result was a loss to Japan of three-quarters of its foreign trade and 90 percent of its oil supplies. This created a crisis. The last civilian prime minister of Japan resigned on October 16, 1941, and was replaced as head of government by General Hideki Tojo.

Shortly Tojo offered to the United States a compromise in which Japan would withdraw from Indochina and recognize an Open Door policy in China, in exchange for America and its allies ending the embargo. Secretary of State Cordell Hull demanded also that Japan sever its ties with Germany. This seemed possible, since the Japan-Germany alliance had never accrued much to Japan's benefit.

Hitler worried that a Japanese–United States modus vivendi would free the American Navy from having to guard two oceans at once and enable it to shift its ships into the Atlantic to counter Germany's. He also worried about the untenable situation of his forces approaching Moscow. So Hitler offered to sign a pact with Japan guaranteeing that the Third Reich would declare war on the United States if Japan did, an offer he hoped would encourage Japan to fight the United States and the Soviet Union for him. The British, for their part, were concerned that the United States and Japan might sign an agreement that would free Japan to attack Singapore and other British possessions while it prevented the United States from coming to the aid of British colonies.

This diplomatic tangle was made worse by Western cryptographers' inability to discern Japan's true intentions. A new Japanese military code had been introduced in the summer of 1941, and by October, American cryptographers had not finished deciphering it. But they were able to report to military superiors an unusual amount of radio traffic indicating a significant military buildup by the Japanese in the eastern Pacific. In late November 1941, with the diplomatic impasse still unresolved, Washington used partially decrypted messages as the basis for sending word to American installations in the Pacific warning that a surprise attack could be imminent; "Hostile action possible at any moment," General Douglas MacArthur in the Philippines was told. His own cryptographers on Corregidor also sounded warnings based on their decodings, which were, however, also still incomplete. MacArthur and other American commanders recognized that the Philippines, Wake, and Guam could be the target of a Japanese bombing attack, and they prepared for it. But as late as December 6, 1941, the American naval base at Pearl Harbor was believed—on no scientific evidence—to be well out of the range of Japanese bombers, even those launched from carriers.

The Japanese had also misconstrued American intentions. Recent scholarship has revealed that in the fall of 1941 the Japanese had broken the American and British diplomatic codes, and that information from the intercepted U.S. diplomatic cables of late November gave the Japanese war cabinet reason to believe that President Franklin D. Roosevelt was taking such a hard line in regard to China and Indochina because the U.S. had decided to go to war. This mistaken belief—the result of misinterpretation of the cables—evidently prompted the Tokyo government to send the attack signal to its forces already steaming toward Hawaii.[1]

On the morning of December 7, 1941, well before the attack on Pearl Harbor, cryptanalysts in Washington, D.C., deciphered Japanese messages that would have warned of it, and radar screens in Hawaii detected Japanese planes almost an hour before they were in bombing range of the harbor, but through a combination of human error, arrogance, interservice protocol, and equipment failure, the warnings were not transmitted and acted upon in time.

[1]Valerie Reitman, "Japan Broke U.S. Code Before Pearl Harbor, Researcher Finds," *Los Angeles Times*, December 7, 2001.

SECTION THREE

World in Flames

| CHAPTER SEVEN |

1942, Year of Trials

IN ADDITION TO destroying a substantial portion of the American fleet and land-based planes at Pearl Harbor on December 7, 1941, the Japanese also decimated American planes at bases in the Philippines, Guam, Midway, and Wake, and at British airfields in Hong Kong and Singapore, while Japanese troops landed at Bataan in the Philippines and on Malaya and the Dutch East Indies, and seized American garrisons at Shanghai and Tientsin. Shortly Japanese planes sank the British Navy's *Repulse* and *Prince of Wales*.[1] The second worldwide conflict of the century had already begun in earnest when Great Britain and the United States declared war on Japan. For four days, Germany took no diplomatic action, providing a faint hope that Hitler would not honor his promise to Japan. Then Hitler

[1]American Naval officers believed that one reason these two British ships were so easily sunk was that they lacked adequate antiaircraft guns, such as were then installed on American capital ships.

did so, and the United States widened its declaration to include Germany and Italy. The Soviet Union did not declare war on Japan, with which it shared a nonaggression pact. The United States, as well as the other Allies, was now engaged in an enormous, two-ocean conflict in which victory seemed at best to lie far in the future.

Dr. John Moorhead, who had developed a specialty in wounds and battlefield injuries during the Great War, had been scheduled to give a lecture to the medical community of Pearl Harbor on December 7; hearing the sirens, Moorhead went to the hospital and started saving lives. He and the hundreds of other medical personnel in Hawaii quickly discovered the limitations of giving morphine injections for pain and saline solution to replace blood. Hospitals were overwhelmed, mandating triage procedures to sequence treatment of the injured. Tannic-acid jelly, the most readily available treatment for burns, had to be put into flit guns, of the sort used for insecticide, to spray onto the many victims pulled from the water or from burning ships. Pearl Harbor provided as rude an awakening to American medicine as it did to the American military.

The medical shocks continued as American and Philippine troops on Bataan mounted a defense against the Japanese—and against malaria, the mosquito-borne disease that debilitated soldiers even when it did not kill them. There was no cure for malaria, and quinine to ameliorate its symptoms was in very short supply. Soldiers weakened by malaria also fell prey to diseases brought on by poor diet. As the defenders' food supply was reduced to a thousand calories a day, the age-old scourges of scurvy, pellagra, beriberi, and amoebic and bacillic dysentery also ran through them. Vitamin deficiency impaired the healing of wounds, and gas gangrene made it necessary to amputate limbs that in more sanitary circumstances could have been saved. General Douglas MacArthur attributed the 1942 defeat of American and Philippine forces on Bataan to disease, and while that judgment has since been questioned—because the Japanese forces were equally riddled with malaria and the other diseases—a basic truth came to the fore: in the theater of the Pacific, "General Disease" would be as great an influence as "General Winter" was on the Russian front.

In one small way the end of the fighting on Bataan was fortunate for the American and Philippine defenders, because had the stalemate continued, the Japanese military planned to release plague-infected fleas against the defenders. Ten attacks were scheduled, part of a larger motion to use biological weapons against targets in China, Samoa, Australia, India, and Alaska. Prototype anthrax bombs were being tested by the Japanese in parts of China, where the Japanese had already used gas and practiced dropping bacteria bombs. Reports of experiments on human subjects done by Major Ishii's group filtered through to the Japanese scientific establishment in papers that substituted the term "Manchurian monkeys" for the Chinese peasant subjects. Many people in the medical community of Japan also knew firsthand about Ishii's CBW work from his annual lectures to medical-school classes, and members of the Imperial household were also informed about it through cousins of the emperor who volunteered to work on the projects.

As the Japanese conquest of the Far East continued, easily overwhelming Dutch, English, Australian, American, and indigenous military forces and bringing many new lands under Japanese control, a Japan Technology Board was announced; among its objectives were spreading the "Japanese character of science and technology" to newly conquered territories and better use of the "resources in the Greater East Asia Co-Prosperity Sphere" for scientific purposes. Confident of empire, Japan shortchanged immediate war-related work in favor of projects for the far future. Most of the twenty-one new research institutes chartered were within private industry, and the Technology Board's first announced goals were in aeronautics: research on supersize aircraft and on very-high-speed, high-altitude planes that were not expected to fly for several more years. And those planes' goal seemed commercial, not military: to carry 500 passengers and 40 crew from Osaka to Surabaya (in the Dutch East Indies) by way of Manila, in fifteen hours. Some military-related work was funded at this time, such as that of Sin-itiro Tomonaga, a former student of Heisenberg who had returned to Japan and was working on microwave radar. The Technology Board allocated almost as much money to build a new cyclotron as to radar research, in the hope, as one of its reports put it, that the cyclotron could lead to "the invention of surprise attack weapons" based on uranium.

Hitler was determined to subjugate the entire Soviet Union in 1942. That January, in the service of that objective, doctors, nurses, and technicians from the "T-4" euthanasia operation in Berlin reached the Russian front. Experienced from having executed 100,000 of the retarded, the physically handicapped, and the senile, they now undertook the mercy killing of German soldiers whose wounds were so severe that they would never be able to return to normal lives. So many of the wounded had limbs frozen on the battlefields and in the hospitals or had been allowed to become gangrenous, which led to amputated limbs and disfigurement, that there had been complaints from relatives at home. The mercy killings were done to quiet these complaints and also to satisfy the military's need not to allocate equipment and personnel to care for these wounded or to transport them home during the spring offensive against Russia.

The accomplishments of the T-4 teams figured in the deliberations of the January 20, 1942, conference of Third Reich bureaucrats at Wannsee, outside Berlin, on the "final solution" to the Jewish "question." Amalgamating bureaucracy and technology, the conferees ratified a plan that had been in gestation since Göring had directed them six months earlier to devise methods and assign priorities to the uprooting, transport, and killing of Jews from Germany

Left to right: physicists Samuel Goudsmit, Clarence Yokum, Werner Heisenberg, Enrico Fermi, and Edward Henry Kraus, at Ann Arbor, Michigan, in 1937.

Before the war, the world's physicists were friends. Near the war's end, Goudsmit was the American officer charged with finding out whether Nazi Germany was close to making an atomic bomb.

Goudsmit's team sifts through the wreckage of the nuclear laboratory led by Heisenberg at Berlin-Dahlem. Items found included blocks of uranium oxide, graphite, and lead.

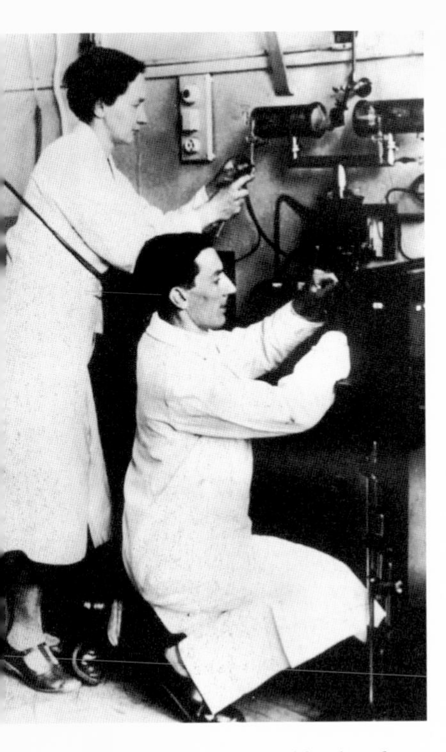

Irène Joliot-Curie and husband
Frédéric Joliot-Curie.

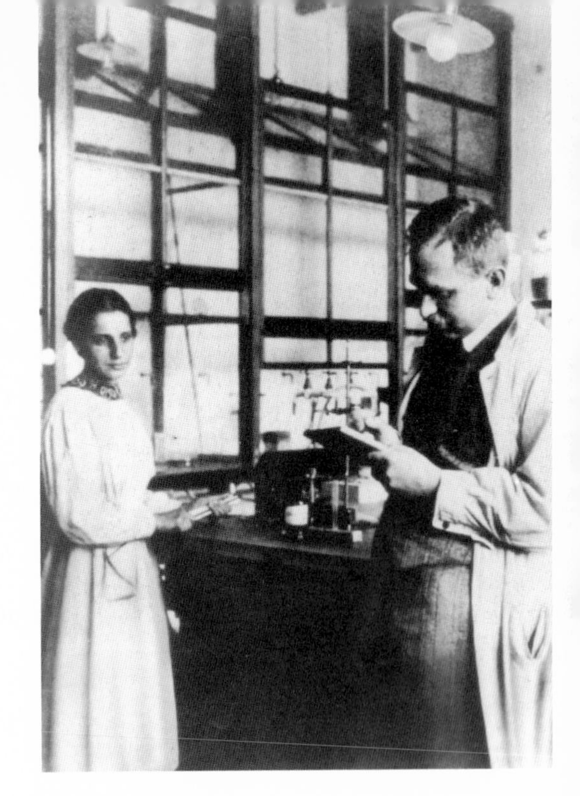

Lise Meitner and Otto Hahn.

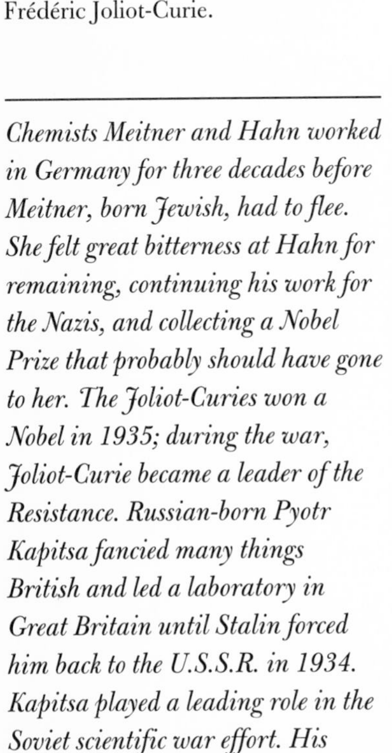

Chemists Meitner and Hahn worked in Germany for three decades before Meitner, born Jewish, had to flee. She felt great bitterness at Hahn for remaining, continuing his work for the Nazis, and collecting a Nobel Prize that probably should have gone to her. The Joliot-Curies won a Nobel in 1935; during the war, Joliot-Curie became a leader of the Resistance. Russian-born Pyotr Kapitsa fancied many things British and led a laboratory in Great Britain until Stalin forced him back to the U.S.S.R. in 1934. Kapitsa played a leading role in the Soviet scientific war effort. His Nobel was delayed until 1978.

Pyotr Kapitsa.

Left to right:
E.T.S. Walton, Sir Ernest Rutherford, John Cockcroft.

Henry Tizard.

Prime Minister Churchill, Frederic Lindemann (in suit), and others watch a new weaponry demonstration on June 18, 1941.

Seared by the loss of young scientists at the front lines in World War I, Rutherford pushed for a registry of scientific talent that could be mobilized in the laboratories to fight the next war. Protégé Walton, who would win a Nobel, did not serve in World War II, but protégé Cockcroft, who shared Walton's Nobel, became a major figure in the war. Lindemann's long battle with Henry Tizard for control of the scientific war apparatus only tilted in his direction when his patron, Winston Churchill, became First Lord of the Admiralty.

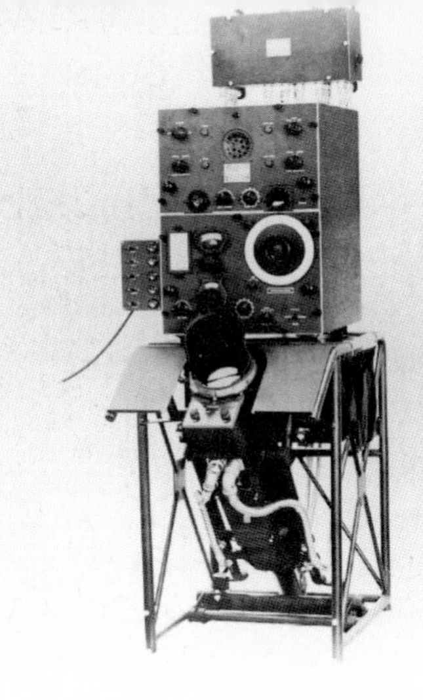

High Frequency Direction Finding (HFDF) radar in "umbrella" mast display aboard an Allied ship in the North Atlantic; and its control unit.

Steady work on Great Britain's scientific and technological war readiness, especially the development of radar, was the key to the British Isles remaining unconquered during the war's early years.

Ground-based air-intercept radar in use by Allied forces in North Africa, 1943.

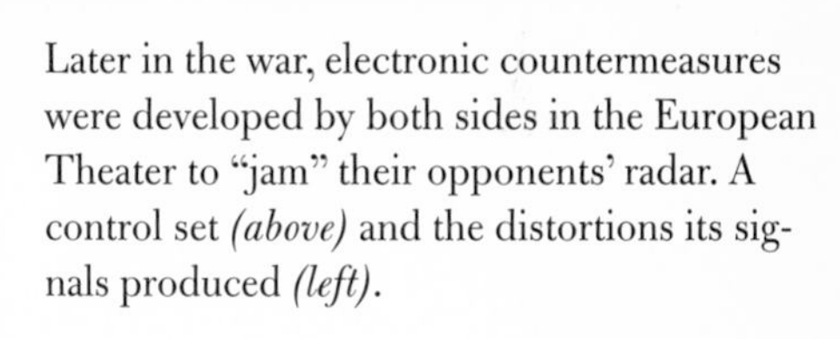

Later in the war, electronic countermeasures were developed by both sides in the European Theater to "jam" their opponents' radar. A control set *(above)* and the distortions its signals produced *(left)*.

NO JAMMING

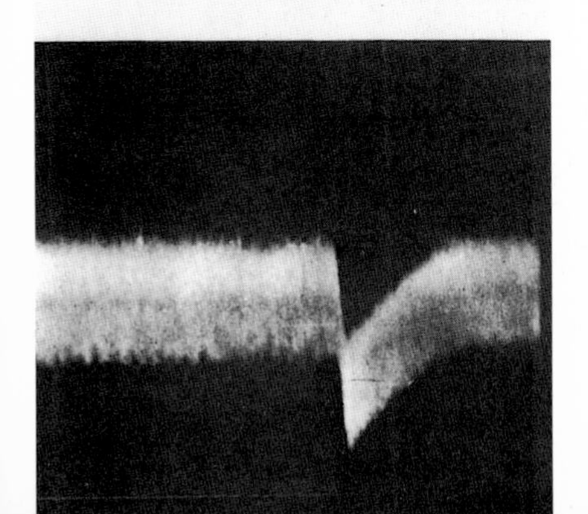

Radar took many forms. "Huff-duff" microwave radars located enemy surface and air vessels and could not be traced by German technology.

Vannevar Bush, leader of the U.S. National Defense Research Council and, later, of the Office of Scientific Research and Development.

Sorting the cards of the U.S. National Register of Scientific and Specialized Personnel.

Bush's efforts to begin the marshaling and directing of American science-based resources toward war, between 1939 and the Japanese attack on Pearl Harbor in late 1941, were backed by President Franklin D. Roosevelt. Manpower resources were rapidly routed to appropriate projects by means of the National Register of Scientific and Specialized Personnel. Admiral King at first refused to allow Bush's scientists to assist in the Navy's antisubmarine efforts but later became a champion of science and technology working with the military.

Below: President Franklin D. Roosevelt, flanked by General George C. Marshall *(left)* and Admiral Ernest J. King *(right)*. Memorial Day parade, May 30, 1942.

Japanese atomic physicists Shohichi Sakata *(left)*, Hideki Yukawa *(right)*, and Nobelist Shin-ichiro Tomonaga, photographed during the 1960s.

Leading scientists in Japan for the most part avoided using their talents for the Japanese war effort. In the United States, where thousands of Japanese-Americans were interned during the war, a few were asked by Cal Tech chemist Emerson to use their expertise to grow guayule as a substitute for natural rubber from the Far East.

Robert Emerson of the California Institute of Technology *(extreme right)* confers with Japanese-American internees at the Manzanar Relocation Camp in northern California.

Depth charges from U.S. Coast Guard cutter *Spencer* explodes German submarine U-175 on April 17, 1943.

A combination of science-based efforts—radar, sonar, Operations Research mathematical analysis, shore-based submarine-spotter planes, and better depth charges—increased the effectiveness of Allied antisubmarine warfare and turned the tide of the Battle of the Atlantic.

Penicillin was discovered in Great Britain, but when British firms were unable to make the antibacterial agent in quantity, the task was given to the U.S. Department of Agriculture, whose Northern Regional Research Laboratory, in Peoria, Illinois, developed a process for manufacturing enough penicillin. The "wonder drug" had a dramatic impact on the recovery rates of wounded and infected military personnel. In earlier wars, more people had been killed by typhus and malaria than in combat. Allied medical and pest-control innovations kept typhus to a minimum and eased the ravages of malaria.

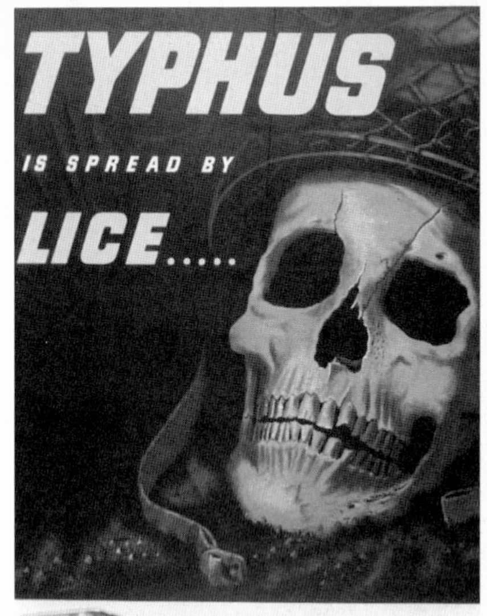

Poster warning of the dangers of typhus.

Above: Penicillin being fed a mixture of corn steep liquor, inorganic salts, and sugar by A. J. Hoyer of the U.S. Department of Agriculture, August 1943; and *(right)* large-batch "fermentation" production of penicillin.

A V-1 robot romb courses above London.

The long-barreled guns dotting the Normandy coast were to have been supplanted by the Nazis' still more ambitious designs for terror weapons, but the Allies bombed them into silence. The invading Allies and resistance fighters found few V-1 launchers, as the retreating Germans had removed them from their concealed mobile platforms.

A V-2 on its launching pad at Peenemünde.

Long-barreled gun capable of shelling Great Britain from France.

The robotic V-1 rocket-propelled bombs, launched from mobile platforms or rail-like tracks that were easily concealed, wreaked havoc over London, until matched by a combination of technologically sophisticated aiming and firing devices and proximity-fuze shells. There were no possible defenses for the Allies against the Germans' enormous V-2 rockets. By the war's end, the Germans were developing a fusion of rocket technology and long-barreled guns for a V-3 weapon.

Launch site of V-1s, near Dieppe, in France.

Walter Dornberger *(left)*, Werner von Braun *(center)* (with arm in cast), and other German rocket scientists, as they were captured in Austria by the U.S. Army on May 3, 1945.

Japanese rubberized-paper balloon, manufactured by drugged schoolgirls and sent across the Pacific carrying a bomb underneath. This one reached Montana.

In contrast to the German V-2s created by Dornberg, Von Braun, and their cohorts, Japan's vengeance weapons were crude: kamikaze suicide planes, which caused thousands of deaths, and balloons that floated on upper atmosphere currents to the North American continent, which killed a few people.

"Bat" guided missile being released, 1945.

By 1945, the prototypes of future weapons systems had been produced. The "Bat" missile-torpedo, seen here being released from a plane, sent out radar pulses and used their reflections to home in on its targets, ships in Japanese-controlled territory. Many German models of advanced weaponry were produced but never reached the battlefield. This experimental pilotless plane was found by advancing Allied troops in a German hangar.

Experimental German guided missile plane, in hangar, 1945.

The atomic bomb explosion over Nagasaki, as seen from a U.S. Air Force plane.

The culmination of billions of dollars and countless scientific man-hours, the atomic bombs dropped on Hiroshima and Nagasaki in August 1945 were said to have ended the war, not won it.

Damage from the atomic bomb in Hiroshima.

and the occupied lands.[2] The Wannsee participants decided to concentrate the efforts in five death camps, most sited in Poland, away from prying eyes; and in order not to arouse suspicion in the victims, to tell the Jews they were being sent east to be resettled.

To increase the efficiency of the exterminations, the participants determined to use improved methods of mass killing. As the SS had learned after machine-gunning large numbers of people, even the most eager of executioners could kill and bury only a few hundred or a thousand people a day without becoming exhausted and without the bodies of the victims taking up a great deal of land area. But in the previous month death squads had begun to use a more efficient technology: three specially constructed mobile vans that suffused lethal gas into the interior compartment to execute passengers as they were being driven about; in two months, 97,000 people had been killed in the vans. The Germans at Wannsee planned to reproduce the gas-van technology on a stupendous scale. Gas-chamber buildings would resemble communal showers so that the victims would enter them without undue resistance. A linked system of ovens would cremate the dead and reduce their volume. The aim was for each camp to be capable of killing several thousand people a day, a total of millions each year. The certainty that this technology would be deadly efficient underlay Hitler's broadcast speech on January 30, 1942, in which he promised that "the result of this war will be the complete annihilation of the Jews."

One week later an important event for German science occurred. After an audience with Hitler at the Russian front, Dr. Fritz Todt, minister of armaments and munitions, boarded a plane for Berlin. Shortly after takeoff, Todt's plane crashed, killing a minister, Albert Speer later wrote, to whom Hitler had paid "a respect bordering on reverence." In addition to being chief of armaments, Todt had been in charge of road building, navigable waterways, power plants, and con-

[2]Zyklon-B had been in "experimental" use in the Birkenau compound at Auschwitz for the previous five months. The first victims had been 500 Russian prisoners of war.

struction throughout the Reich. Moving swiftly to head off a power grab by Göring, Hitler appointed the thirty-six-year-old Speer as successor to Todt in all his various capacities, including the construction of protected bases for U-boats on the French coast, the repair of Norwegian roads, the creation of units for the rocket-building programs at Peenemünde and poison gas factories in Poland, for armaments—and for all of science. Although Speer had no managerial experience, was a stranger to weaponry and science, and was, as he later wrote, "a complete outsider to the army, to the party, and to industry," he was appointed to "one of those three or four ministries on which the existence of the state depended." To appease Göring and to emphasize the importance of science, Speer suggested and Hitler agreed that the marshal become the leader of the Reich Research Council.

Pride as well as British technology suffered a very public defeat at German hands in early February 1942, as the German heavy cruisers *Scharnhorst, Gneisenau,* and *Prinz Eugen* eluded a British blockade at Brest, survived air attacks and hits by mines and torpedoes, and after several days at sea reached Wilhelmshaven intact. The key to the German ships' escape, Reg Jones came to believe, was a technique he recognized from his days as a practical joker—for four days before the cruisers left port, the volume of German radar jamming done by shore-based and airborne transmitters was slowly but steadily escalated, to prevent the British from realizing on the fifth day that the jamming concealed a significant event.

Later in February it was the Germans' turn to be embarrassed. Low-level British aerial reconnaissance had earlier photographed a Würzburg radar station on an isolated point of the French coast known as Bruneval. The station was part of a coastal warning system that alerted the Germans to British air attacks, and Jones wanted to know how it operated so countermeasures could be designed to neutralize it. A daring parachute and naval raid on Bruneval overcame many obstacles and mistakes—including dropping the parachutists more than a mile from the target—to bring back intact the working parts of the station along with a German technician who knew how to operate it, a cargo almost beyond value. From the fruits of this single stroke, the British were able to construct a jamming technique for the most effective of German radars.

Two years after the Nazis had overrun Western Europe, Resistance units were well entrenched among the French, the Belgians, and the Dutch, but when discovered, they often suffered grievous losses. In February 1942 Langevin's son-in-law, Jacques Solomon, a Jewish physicist who worked with Joliot in the underground, was executed by the Nazis for his Resistance work, and Langevin's daughter, along with Solomon's mother, was sent to Auschwitz. After that, Joliot became much more active in the Resistance: he joined the Communist Party, which was the best-organized Resistance group, and became president of the Directing Committee of the National Front, the leading organization. He managed to hide these activities from the Nazi authorities, but Gentner figured them out. As Gentner later recalled, he told Joliot, "I do not like to know what you are doing, because this is dangerous, if I know." Also, when Joliot would be arrested on suspicion of such activities, Gentner would spring him from jail; Gentner was able to do so "because I knew somebody [and] because I told my office that . . . cross-sections are extremely important to the German victory, and if you leave Joliot in the prison, the workshop and all the technicians don't work." If Gentner had not intervened, Joliot would doubtless have suffered the fate of Solomon or of three French women scientists, among them one formerly associated with the Curie Institute, who were executed by the Nazis that spring for "De Gaullist" activities.

When Reg Jones learned of a new radio-navigation beam system located in northern France that guided German night-fighter planes, he asked the Resistance to obtain more details. He received a report whose technical sophistication astonished him. It was compiled by French physicist Yves Rocard after a surreptitious visit to the beam station. British Intelligence then extracted Rocard from France for debriefing, but because they did not return him quickly enough to avoid suspicion about his absence, Rocard had to remain in Great Britain, where he became director of scientific research for the Free French Navy.

In the overrun countries, Allied Intelligence also relied on local people who were not members of the Resistance. Jones dropped pigeons in delayed-release crates, accompanied by questionnaires that those who found the crates could fill out to indicate the presence of nearby radar installations, questionnaires that the pigeons would then bring back to their British handlers. Information from such sources, and from Enigma decrypts, enabled Jones to give the RAF six weeks' notice of a cluster of attacks on British inland cities, later known as the Baedeker raids. Having adequate warning time, the British were able to jam the German navigational beams and lessen the damage done in those raids. Knowledge of the locations of German radar stations, gained from Resistance sources, also enabled the RAF to avoid those stations while making major attacks on Lübeck and Rostock in the spring of 1942.

The statistical analysis of British bombing insisted upon by Lindemann revealed an abysmal record in hitting German targets; to improve accuracy, TRE, the Telecommunications Research Establishment, developed two systems, one of them ready by early 1942. Invented by R. J. Dippy and known as "Gee," it used three separate stations to send synchronized pulses to an aircraft, permitting ground controllers as well as the on-board navigator to precisely determine the craft's position, guide the plane to the target, and release the bombs. In a careless move, the RAF sent bombers with Gee equipment over Germany while Gee was still in prototype; one of the planes was shot down and its equipment captured. Jones was asked to dissuade the Germans from suspecting they had captured an important device. He abolished the name "Gee" in documents and substituted one that had to do with an advanced telephony system, camouflaged the Gee sending stations physically to look like ordinary radar installations, and provided a substitute focus for German inquiry—another bombing-control system deliberately labeled "Jay" because German speakers had difficulty distinguishing between J and G. He also staged a scene at the Savoy Hotel in London, making sure that a known German agent overheard a conversation about the new secret system called Jay. Transmitted to Germany, this false information was considered more credible than any other because it was obtained by deceit.

Another British locating system was called "H_2S"—because when it was originally conceived, its developers thought that its potential stank and gave it the chemical symbol for the rotten-egg smell of hydrogen sulfide. It scanned the ground below the aircraft, displaying on the cathode-ray tube a picture that could be cross-checked with charts to provide very accurate readings of the ground. The development of H_2S suffered a setback in the spring of 1942 when a bomber on which a prototype was being tested crashed in British territory, killing five scientists who were leading the project. The crash also occasioned further delay because British officials became concerned that if a bomber with H_2S went down over Germany, the cavity magnetron that powered the system might be discovered, compromising this highly guarded secret. Accordingly, the authorities substituted another power source for the cavity magnetron in the H_2S. This caution was well founded: when a bomber carrying H_2S did crash in German-controlled territory near Rotterdam, the set was given to General Wolfgang Martini of the Luftwaffe, head of the Signals Division since 1933. Within weeks Martini's crew had rebuilt the H_2S equipment. Damaged again in an Allied raid, the precious set was transferred to Berlin and rebuilt once more. It worked very well, providing from a rooftop some "excellent images of the Berlin residential area."

Later on in 1942, while Hitler was at Kalinovka in southern Russia, he sent for General Martini and in the presence of Göring berated Martini over the progress of radio-wave technology. Hitler recounted how he had set out on a flight to southern Germany, directed by shortwave radar but due to navigational mistakes had ended up in northern Germany. Failure of his own radar led Hitler to question Martini's assertion that the British were able to bomb specific munitions factories at night or during cloudy weather. When Martini tried to explain the technical possibilities, Hitler kept asking questions until the general became confused. Upon returning to his regular post, however, Martini found that Hitler had assigned radar a new and more important priority. One of Martini's first actions was to ask for the recall from active duty of engineers and technicians who had once worked on centrimetric radar. They used the captured H_2S set to construct radar detectors of microwave and H_2S transmissions from Allied ships and planes.

By then the Raytheon Corporation and other American commercial companies were producing 25,000 cavity magnetrons per week. If the NDRC, and then the OSRD, had not been well established in the United States prior to Pearl Harbor, it is unlikely that American science and technology would have contributed very much to the Allied effort throughout the rest of the war.

Japan's conquests cut off the United States' supply of natural rubber—even in peacetime the United States used half the world's annual production—and there were also shortages of gasoline products. Large-scale manufacture of butadiene and styrene synthetic rubbers began, but there were no further scientific breakthroughs on rubber.

A program to grow guayule, a common shrub in Mexico with rubber-like properties, began in an odd location, the Manzanar Relocation Camp in California. Caltech chemistry professor Robert Emerson sought out former colleagues at the university such as Dr.

Shimpe Nishimura, and other Japanese-American internees, to work on the project. The internees managed to hybridize the plant, and create a new manufacturing process that produced rubber of greater tensile strength than any other. Guayule was used to line fuel tanks to make them re-close after punctures by bullets, but it was never manufactured in large enough quantities to replace synthetic rubber (made from oil) in other uses like tires.

Although political maneuvering by the oil companies succeeded in preventing the large-scale manufacture of guayule rubber, oil-industry companies that had formerly been rivals to one another now allowed their chemists to collaborate to find ways to manufacture higher-octane gasoline for aviation and lubricants, and toluene for TNT. Research institutes like the Mellon in Pittsburgh put aside other work to develop synthetic antimalarial substitutes for quinine, whose supply also had been cut off.

Inventing new technologies was much chancier than improving existing ones, and in the latter practice American scientists and technologists excelled, bringing their already fabled improvisational skills to bear on leads provided by the British. A most important instance: the proximity fuze. In theory this was a simple device, a transmitter and receiver located in a projectile, acting as a fuze to detonate an explosive projectile only when it was in proximity to the target. In the 1930s the idea of such a fuze occurred independently to inventors in Germany, the United States, Poland, and Great Britain. But it was a long way from conception to production, and if a proximity fuze was improperly designed, it might explode closer to the gun that fired it than to the target, becoming, as one American naval officer suggested, "the world's most complicated form of self-destroying ammunition."

The Tizard Mission had carried an early British version of the proximity fuze to the United States, where the task of developing it was given to Merle Tuve, described by peers and subordinates as "the most dynamic man I ever met." Born in Canton, Ohio, Tuve had been a childhood friend of his neighbor Ernest Lawrence; the two boys constructed a telegraph between their homes and later communicated by ham radio. They pursued parallel careers in the sciences, Tuve moving from chemistry to electrical engineering to practical physics. He had pioneered the measurement of the upper atmosphere by radio waves and had also been one of the first Americans to experiment with the products of nuclear fission. He was an early member of the Uranium Committee but decided that rather than work on what became the Manhattan Project, he would devote himself to a weapon of more immediate practicality: the proximity fuze.

As the NDRC was formed, Tuve became head of its "Section T," which was shortly attached to the Johns Hopkins University and moved to a converted garage in Silver Spring, Maryland. Among Tuve's immediate tasks was to determine which sort of trigger to use in a proximity fuze. By the early spring of 1942 the promise of photoelectric or acoustic triggers had been eclipsed by the subtlety and variety offered by a radio device. Tuve saw his mission as turning theory into a reliable radar-triggered proximity fuze that could be mass-manufactured. That meant first solving such problems as providing a battery power source—one that would not decay while the fuze was on the shelf or being transported to the battlefield—and reducing the size of the trigger to fit inside the head of an artillery shell, rather than, as with the British fuzes, into the much roomier head of a small rocket.

Germans and Japanese were also trying to solve those fuze problems. The German work was under Dornberger and his rocketry team, and the Japanese effort, at the nuclear research laboratory under Nishina. As Dornberger later wrote, although thirty different designs were made and tested, "It proved impossible, throughout the war, to get such a device manufactured in Germany." We will note later what the Japanese effort was able to produce.

Tuve's significant success with the American proximity fuze owed more than a little to his style, articulated in a set of rules for his co-workers, such as "I don't want any damn fool in this laboratory to save money, I only want him to save *time*"; "Run your bets in *parallel,* not in *series;* this is a *war* program, not a scientific program"; and "The *final* result is the only thing that counts, and the only criterion is, Does it work *then.*" He also made clear that the scientist must be continually accountable for his war work, insisting that "our moral responsibility goes all the way to the final battle use of this [proximity fuze]; its failure there is our failure regardless of who is technically responsible for the causes of failure."

Tuve later came to believe that his group's eventual success was a function of their having practiced democracy in the laboratory: Anyone could, and frequently did, contribute to the effort by making suggestions on the directions to go in, how to improve this or that factor. Tuve knew from his wide experience and acquaintance with German researchers that in the labs of the Reich, there was no democracy: Every individual knew his place in the hierarchy, and those with low rank and low status were not invited or permitted to contribute to the design of the work. In later years, Tuve in an interview would cite this fundamental difference as the key to why German laboratories "kept on making the same dumb mistakes all through the war."

The first test of a radio-equipped fuze was in April 1942, for a group of Navy observers and Vannevar Bush; shells fired from a battleship shot down three airborne drones. Bush sent a telegram to Conant: THREE RUNS, THREE HITS, NO ERRORS. The Navy set stringent requirements for manufacture: the fuzes would have to be built by ordinary factory hands, not by scientists, and would have to work 52 percent of the time before the Navy would award a contract to produce them. When this reasonable goal was reached, production began, 400 a day at first, escalating eventually to 70,000 per day.

While the proximity fuze was being developed, hundreds of similar scientific and technical problems dealing with other pieces of military equipment were being addressed by American university-based and commercial-manufacturer research groups. For instance, early combat had exposed glitches in the operation of optical range finders. Princeton researchers discovered that some of the errors came from unexpected variations in air pressure in the instrument; they found a way to use pressurized helium instead of air inside the instrument so it could work properly no matter what the surrounding temperature or outside air pressure. At Eastman Kodak the problem tackled was a distortion in the range finder's view of the target. A process called "autocollination" arranged the optical system so that light rays from the reading scale passed through optical elements and became subject to the same mechanical shifts as did rays from the target, eliminating the errors.

In the spring of 1942 the most important advanced technology being readied for the battlefield by American research was a "predictor" that rapidly computed the likely path of an incoming plane so that antiaircraft fire could be directed at it. Warren Weaver, the former Rockefeller Foundation executive who had become the leader of an OSRD section, had accompanied Conant on his trip to London, where he and associate Edward J. Poitras learned firsthand the terror and frustration of not being able to shoot down incoming planes with antiaircraft fire—at least 10,000 rounds were being expended for each plane brought down by ground-based guns. Returning to the United States, Weaver and Poitras looked for ways to improve that ratio. First, however, good data had to be obtained. Precisely how accurate was antiaircraft fire? How could you best measure accuracy? What were the sources of the lack of accuracy? Individual questions were portioned out to Bell Labs, Kodak, MIT, Caltech, Princeton, Tufts: work on the "psycho-physiological" errors made by humans operating ack-ack guns; on the possibility of replacing the human eye and hand with mechanical "servomechanisms"; on basic optics; on the theory of computers; on the theory of prediction; on "dead time"—the moments when a shell was being transferred from the fuse-setting machinery into the breech of the gun. Their findings translated into a conclusion that the best way to direct antiaircraft fire to the right location was to replace the human eye and hand with computer-assisted machinery.

The central element of the best design came to engineer David Parkinson in a dream: he had previously invented a potentiometer that controlled the motion of a pen to automatically record changes in voltage in a phone system, and one night he dreamed that his potentiometer was similarly controlling the fire from an antiaircraft gun.

Information gathered from many laboratories was synthesized by Bell Labs and the MIT Rad Lab, which designed a prototype predictor system ready for testing in April 1942. It worked so well that the next day the Army ordered the first 1,256 units of the SCR-584 tracking radar and the connected M-9 computerized predictor. Shortly the services requisitioned several hundred million dollars' worth of the predictors and directors.

Three other technologies nearing full development at this time at MIT owed their existence to Luis Alvarez, a Lawrence protégé and rising star in physics who had been working at the Rad Lab since its inception. Two devolved from Alvarez's interaction with Taffy Bowen, late in 1941: a precision-bombing radar set and a Microwave Early Warning radar. The third was a blind-landing system. An early attempt at such a system failed in trials, and Alvarez was greatly discouraged. He and Loomis met, and when Alvarez started to commiserate about the failure, Loomis challenged him to solve the problem before the night was out. Alvarez did. Years later, Alvarez wrote that "without Loomis's ultimatum that night, World War II would have seen no effective blind-landing system. I would have immersed myself in other interesting projects to forget my disappointment and embarrassment. Many lives would have been lost unnecessarily."

In his work on the various bombing devices, Alvarez had occasion to examine the vaunted Norden bombsight, and found a fundamental defect. For it to operate properly, the airplane carrying it had to fly a fixed course for several minutes before a bomb was to be released. You could keep to a fixed course during peacetime, Alvarez learned, but to do so on a bombing run over Germany's devastating antiaircraft batteries would be "distinctly unhealthy." So, even though it could be accurate, the Norden bombsight was almost useless in combat. To replaced the Norden bombsight, Alvarez helped design an offshoot of the blind landing system, a blind bombing system called EAGLE. It did not make its appearance in battle until late in the war, but on its first missions it achieved a 95 percent destruction rate, hitting virtually every target at which it was aimed.

Canadian scientists almost single-handedly devised a safe production process for the very powerful but highly unstable chemical explosive RDX and did significant work on de-icing airplanes, designing vehicles that could move through snowy landscapes, and aerial photography. But what truly earned Canada a position at the table where the scientific and strategic direction of the war were planned was its work in chemical and biological warfare. The vast open areas of the Canadian west were the attraction, as the United States and Great Britain readily agreed to test chemical and biological agents there for the benefit of all the Allies or, as they were now called, the United Nations. Western scientists and military planners had an underlying, horrific fear of a "gas Pearl Harbor," in which Japan or Germany—both known to have stockpiles of poisonous gases—would mount a sneak attack against London or a North American city; that fear was accompanied by the belief that the only sufficient defense against gas attack would be equivalent Allied stockpiles that could be used in retaliation. In what C. J. Mackenzie labeled "the most expensive experiment ever performed in Canada," 95 tons of phosgene gas were blown up by dynamite, to test whether gas masks could withstand a very large cloud, one that spread for 17 miles. After this large-scale experiment there was greater cooperation among the chemical-warfare services of Canada, the United States, and Great Britain.

A leading Canadian scientist/administrator and his assistant used to play chess over lunch—without a board. A similar but far deadlier game was being waged between the Allied and Axis experts in chemical warfare. When information came to the Allies that the Germans were moving around trainloads of certain chemicals, Allied chemists deduced that the Germans were making cyanogen gases. No extant gas mask would protect a soldier from cyanogen, but shortly a charcoal-plus-chromium combination was devised in Canada that would neutralize it. Knowledge of the chemical composition of the neutralizer was kept secret, and the Germans then had to wrestle with the possibility that while the Allies had a gas mask that neutralized cyanogen, they did not. Conversely, Allied scientists and military planners had many sleepless nights upon learning that they could not counter the new German nerve gases.

Biological warfare was already being used by several of the Allies, albeit on a relatively small scale. The Polish liaison officer in London would later admit that Polish partisans in Warsaw had poisoned or assassinated hundreds of Germans using typhus, typhoid, and arsenic. In 1942 Soviet partisans were killing Germans by spreading arsenic powder. In May 1942 British-trained Czechoslovakian partisans assassinated Hitler's favorite, Heinrich Heydrich, the chief of the security forces, in Prague, with a grenade containing a virulent strain of botulism developed at Britain's Porton Downs. Fortunately for the rest of the world, the Nazis never figured out what had killed Heydrich, or they might well have replied in kind; as it was, in reprisals for Heydrich's death the Nazis accelerated the roundup and execution of 10,000 Czech Jews. And in mid-1942 teams from Porton Downs exploded an anthrax bomb on Guinard Island, off northern Scotland, killing all the sheep and making the island permanently uninhabitable.

In the late spring of 1942 American poet Archibald MacLeish, then director of the government's Office of Facts and Figures, delivered an address to the American Library Association in which he clearly identified what was at stake in the scientific war. MacLeish pointed out that the Nazi regime had consistently and in various ways attacked freedom of thought, of investigation, and of speech—individual "rights" that were at the heart of all scientific research. The Nazis had attempted to eliminate the intellectual leadership in every country they conquered, sacked libraries and looted scientific institutes, degraded education, and quashed all objections to totalitarian theory. MacLeish worried that American intellectuals had not yet acknowledged that these Nazi depredations amounted to besieging the country of the mind itself. That country, he asserted, could be saved only by courageous counterattack; and he warned that even if battlefield victories against the Nazis were won, the greater war might still be lost should we fail to maintain the primacy and authority of truth and intellectual investigation. To preserve confidence in learning, reason, and truth, intellectuals must throw themselves completely into the task of preserving the West. Allied scientists must no longer be passive or stand on the sidelines and pretend to the need for objectivity. They must take active part in the war effort, because failure to do so would precipitate their countries—and their work—into being swallowed by the darkness of totalitarianism and its utter contempt for the life of the mind.

One of Albert Speer's earliest meetings after his appointment as an important minister was with Albert Vögler, president of the KWG, who in the spring of 1942 called Speer's attention to the "neglected field of nuclear research," in part to emphasize that funding for all basic research had been unduly cut back. Vögler was trying to counteract an army ordnance report that concluded that Germany ought not to mount the large-scale effort necessary to build a bomb. To challenge this conclusion, a tutorial on nuclear research was organized for Speer and for some military leaders.

To this day controversy exists on what was actually said and done at this tutorial. Heisenberg and others who later asserted that Germany never tried to make an atomic bomb, only to build reactors as a power source, contend that at this conference they convinced the authorities to forgo building a bomb because such a feat was not within the capabilities of Germany in the midst of a war. According to Heisenberg, at this conference he, Hahn, and von Laue "falsified the mathematics to avoid the development of the atomic bomb by German scientists."

Albert Speer recollected the day and the decision differently. He recalled being presented with excerpts from American scientific journals in an attempt to prod him to assign larger resources to the atomic program. But when Speer agreed to give the scientists whatever they wanted, he was surprised to learn that they didn't want much. Heisenberg told him that producing a bomb would take a minimum of two years, in part because Europe's only adequate cyclotron was in Paris; when Speer offered to have large cyclotrons built in Germany, Heisenberg said that because Germans lacked experience in building cyclotrons, they would have to start with smaller ones. Speer concluded that "the atom bomb could no longer have any bearing on the course of the war," but he encouraged further research. Speer's version, as does Heisenberg's, implies that the researchers could have made a bomb or created a self-sustaining chain reaction if they had had a few years and adequate resources.

A third version challenges that conclusion. It points out that the leading scientists at the April 1942 meeting expressed no qualms about making a bomb but did advise Speer that an atomic weapon could not be built in the next few years—not by them and not by any other country—because they misunderstood the scientific requirements for making a bomb. They made three basic scientific mistakes: they guessed that the amount of material needed for a critical mass was in tons rather than in pounds;[3] they incorrectly assumed that a U-235 bomb would depend on "fast-neutron" reaction, whose action they did not properly comprehend; and they incorrectly calculated the rate at which neutrons would multiply in a fast-neutron reaction.

[3]American nuclear researchers were just coming to the conclusion in April–May 1942 that the critical mass would be between 2.5 and 5.0 kilograms; an earlier estimate by the National Academy of Sciences had put the critical mass amount at more than 100 kilograms.

Had the nuclear scientists told Speer that their project was feasible but would be expensive—the formula used by Dornberger and von Braun to extract resources for their rocket program—they might well have been awarded a huge budget for development of an atomic bomb, and made one. Speer recalls telling Hitler about the meeting with the nuclear scientists and concluding that his report "strained his [Hitler's] intellectual capacity" while it "confirmed the view that there was not much profit in the matter."

Nuclear research in Germany continued to progress, though, and reaped the benefit of a new and experienced recruit. In Paris the ability of Wolfgang Gentner to protect Joliot from the Germans came to an end; denounced to his superiors by another German member of the group watching the French scientists, Gentner was ordered to return to Germany. There he used his expertise to assist Bothe in the German cyclotron experiments.[4] Gentner's replacement in Paris continued to permit Joliot to operate in peace and prevented Joliot's cyclotron from being moved to Germany.

News that the Germans were making some progress on nuclear research was sent by Soviet espionage agents in Germany to Moscow, where it occasioned a conference among top scientists, who advised Stalin that an atomic bomb was indeed feasible, but that because of the war's disruptions, a Russian bomb could not be completed for another decade. Stalin may have been equally stimulated just then by a letter from a young nuclear physicist. Georgi Flerov had discovered that articles by the leading American, British, and German nuclear researchers had disappeared from print, and deduced that this meant they were working on a bomb, because they were "dogs that did not bark." He told Stalin that by the time Germany, the United States, or Great Britain had succeeded in making a

[4]Like Heisenberg, Hahn, and the German nuclear scientists, both Joliot and Gentner believed that an atomic bomb could not be made during the war, though for different reasons. Bothe's cyclotron, completed in the autumn of 1943 with Gentner's help, was operated only by experimental physicists. The team in charge did not have engineers or theorists, as did cyclotron operations in the United States.

bomb, all that would be left for the USSR to do would be to assign blame. Stalin immediately put atomic bomb research under Lavrenti Beria, head of the secret police, and told him to look for "shortcuts." The next day Beria sent instructions to spies in the United States, Great Britain, and Germany to seek secret information on the development of an atomic bomb.

From this time forward in the Soviet Union, as in the West, there could no longer be any doubt in even the most dedicated Soviet basic researcher's mind that if his project was to have any future, he must first devote his talents and energies toward defeating the Nazis. For Nikolai Semenov this meant translating his studies of molecular flow—relevant to chain reactions—into work on plastics and on the improvement of automobile engines. Theoretical physicists like Igor Tamm, Ilya Frank, and Pavel Cherenkov turned their talents to weapons development. Nikolai Basov and Alseksandr Prokhorov worked on improving the propagation of radio waves for radar. Pyotr Kapitsa's institute began cooperating with Ioffe's toward developing both atomic bombs and nuclear power plants. Kapitsa wrote an article for the *Red Star,* the Soviet army newspaper, extolling the work of mathematicians applying the laws of probability to problems of ballistics and trajectories, botanists using their knowledge of vegetation to design camouflage, physiologists studying the relationship of diet to sharpness of vision, chemists finding a substitute for the oil of balsam that was used widely in treating wounds. The Soviet Central Committee paired specific Academy of Sciences institutes with individual defense agencies, military scientists were put in charge of sections of the Academy devoted to practical war work, and the Central Committee began to exercise day-to-day oversight to ensure that everyone pushed together toward common, war-related goals; all of this resulted in "a far-reaching shift of the national R&D effort toward military needs," Bruce Parrott writes. The involvement of top scientists in technological tasks led to very rapid improvement of airplane aerodynamics, artillery fire patterns, bombs, and military fuels, and it also identified new stocks of natural resources and ways to adapt relocated manufacturing enterprises to the use of local raw materials. Geologic explorations found previously unknown deposits of manganese, aluminum, nickel, wolfram, oil, and coal.

The most immediate boost to the Soviets' ability to resist the German onslaught came from the United States in the form of vast amounts of supplies, military equipment, manufacturing support, and humanitarian aid. To further induce Stalin to keep Hitler's forces occupied on the Russian front—and to placate Stalin for the Allies' not having fulfilled their pledge to open a European front in 1942—ever greater fractions of American Lend-Lease supplies were forwarded by Great Britain to the Soviet Union, even when this meant reduced food rations for the British.

In July 1942 a leading British expert on nuclear physics told Churchill that Great Britain's own research in the field was a "dwindling asset" that would soon be eclipsed by the American work. If Churchill had wanted now to accept the earlier American offer of a nuclear-research partnership, which he had rejected in late 1941, he could not, since the offer had not been renewed once the American effort was in full gallop. The British instead joined forces with the Canadians, sending Joliot's former associate Halban to lead the effort in Canada. Meanwhile, in the United States, control of nuclear research was transferred from the NDRC to the Army and the direction of General Leslie Groves of the Manhattan Engineering District. Vannevar Bush later wrote that he chose the Army rather than the Navy to supervise the bomb project because of his earlier bad experiences with "naval officers, especially those at the Naval Research Laboratory . . . [who lacked] sufficient respect for and an ability to work cooperatively with civilian scientists." The choice, then, was Bush's revenge on the Navy for the difficulties that Admiral Bowen had caused him.

Both British and American nuclear scientists continued to fear that their counterparts in Germany were ahead of them in the race to build an atomic bomb. Samuel Goudsmit later recalled the Allied scientists' reasoning:

> Since the Germans had started their uranium research about two years before us, we figured they must be at least two years ahead of us. They might not have the bomb yet, but they must have had chain reacting piles going for several years. It followed that they must have fearful quantities of artificial radioactive materials available. How simple it would be for them to poison the water and food supplies of our large cities with chemically non-detectable substances and sow death wholesale among us by dreadful invisible radiations.

Enmeshed in such logic, American scientists and managers Bush and Conant could not delay or even question the American nuclear program and the development an atomic bomb to enable the democracies to have a threat in hand to forestall Nazi Germany's use of any similar bomb or radiation products.

In mid-1942 German science began to rid itself of counterproductive elements. A forthright German Physical Society report prepared by Carl Ramsauer contended that German physics had been left behind by American physics in matters of importance to the war effort because of the depredations wrought by Aryan physics. Ludwig Prandtl's emphatic support of Ramsauer's conclusion influenced Göring. In a secret meeting on July 6, 1942, attended by Rust, Vögler, and Nazi Party "philosopher" Alfred Rosenberg, Göring announced that Hitler had asked him to reorganize German science research to concentrate on war-related matters: weaponry, nutrition, and health. The Reich Marshal astonished his listeners by asserting that Hitler now believed that the prior policy of rejecting the research of Jewish scientists had been a mistake; the Reich needed every bit of available expertise, and it would be "crazy" to insist that a brilliant researcher whose wife was Jewish or who himself was half Jewish should be dismissed from his posts and kept from doing research. Hitler, Göring said, had made exceptions to these rules in the past—the chemist Otto Warburg, for instance, who Hitler was convinced would protect him from getting cancer—and now would "make exceptions even more gladly if it is a question of an important research project or researcher."

Ramsauer's efforts at restoring sanity to physics and Prandtl's decade-long attempt to protect his Jewish colleagues were later cited as evidence of the ability of German science to resist the Third Reich, but such efforts did not amount to very much, compared to the active resistance organized in France by the circle around Joliot or by Norwegians willing to participate in planning the destruction of their own heavy-water plant. To German scientists socialized to believe that obedience to state authority and serving the interests of the state were obligations of citizens, acts of sabotage or resistance were construed as treasonous, not as legitimate ways to overthrow a destructive regime. As Helmuth Trischler concludes in a study of resistance in the German scientific community, what was done by Prandtl, Ramsauer, and a few others

> were not feats of resistance aimed at conquering the system. On the contrary, the [scientists] sought to strengthen the powerful position of the Reich and thus to keep National Socialism in power by removing barriers to efficiency—whether ideologically motivated attacks on the sciences by Nazi activists or political clashes of competence by competing powers within the polycratic chaos of the Nazi regime.

A new council was set up, a "presidium" under Göring, for which Speer would oversee research by the military, Rust the universities and KWIs, and Mentzel would build a card index file to avoid duplication of projects and to foster communication and cooperation. One almost immediate result of this realignment was to permit Heisenberg to take a key post in Berlin. The decision became a ratification of the worth of the theorists over that of the Aryan physicists. A second result was to increase the willingness of scientists to do war-related work; Timoféeff-Ressovsky began experiments on radiation-induced injuries, cancer, and the efficacy of gas masks.

There was also a speed-up of the A-4 "big rockets" of Dornberger and von Braun and the competing "flying bomb" project under another arm of the military at Peenemünde. For the first time these programs had the complete backing of the academic research establishment, the armaments minister, and the armed forces, which were beginning to recognize the weapons as potentially decisive for winning the current war—a war that was persisting longer than they had imagined possible.

The Göring takeover reoriented German science toward full mobilization, a posture resembling those already in place in Great Britain, the United States, Canada, Australia, the USSR, and, to a lesser degree, in Japan. Still, proportionally fewer German academic scientists were drafted into military research than were their counterparts in the other belligerent countries, and the Reich and its science administrators continued to take pride that research in certain fields, such as basic biology and genetics, went on relatively uninterrupted by the war.

While the ruling presidium was composed of enthusiastic supporters of science, not all members shared enlightened views. Herbert Backe, a Nazi since 1923, was a member by virtue of being the minister of agriculture, and from that post he did a lot of damage. Backe believed that Russians were genetically inferior, and his program to feed Central and Western Europe involved fostering genocide in the Soviet Union by denying food to its people. There was also some "scientific research" funded by the SS's Institute for Practical Research in Military Science. In grim distortions of the scientific method, Strasbourg anatomy professor August Hirt had 115 Jewish inmates at Auschwitz murdered so he could establish a typology of Jewish skeletons, and at Natzweiler he had dozens of inmates doused with mustard gas so he could measure how long it took them to die. Of Sigmund Rascher's subjecting Dachau inmates to frigid waters, killing 80 of the 300 tested, Himmler wrote, "I consider people who today still reject human experiments, and instead allow brave German soldiers to die from the effects of undercooling, to be guilty of high treason and treason against their country."

Hitler and Himmler visited a death camp together on August 15, 1942; Hitler complained that the pace of killings was too slow, that "the whole operation must be speeded up, considerably speeded up." The remark was noted by an associate because it was unusual: Hitler almost always avoided giving direct instruction on the annihilation of the Jews, leaving that to be done by his surrogates.

The first credible reports of organized, wholesale killing of Jews reached Switzerland that month. At the risk of his life, a German industrialist supplied accurate information to an international Jewish organization, on the death camps in Poland, their lethal gas chambers, and the prussic acid used as the lethal agent.

Prof Lindemann, who in the summer of 1942 was named Lord Cherwell of Oxford,[5] was in adamant agreement with Churchill's early war statement that "bombers alone provide the means of victory," and in their daily meetings—sometimes held well after midnight in the underground bunker at 10 Downing Street—and their weekends together at Chequers, the prime minister's country home, he and Churchill outlined a campaign to carry the war to Germany. Based on analyses by Bernal and Zuckerman of the effect of German bombing on British cities, Cherwell and Air Marshal Sir Arthur Harris

[5]On learning of Lindemann's elevation, Tizard observed that the Cherwell River was a small and muddy stream.

devised a "dehousing" bombing effort whose purpose was to oust Germans from their homes, to demoralize the German populace, and to force Hitler to use the Luftwaffe to defend Germany rather than for other purposes.

Cherwell predicted that the dehousing could be accomplished by mid-1943. Tizard found Cherwell's calculations faulty and he pointed out that in the previous months almost as many British airmen had been lost over Germany as there had been Germans killed by the bombing and that the British airmen were not replaceable. This objection was brushed aside with the comment that Tizard had been listening too closely to Blackett, and the dehousing bombing of Germany went on. Eventually—though not by mid-1943—this bombing would kill 600,000 people and destroy more than a third of Germany's urban housing stock, at a terribly high cost in terms of Allied pilots and planes. In 1942, however, while American and British air commanders believed that the bombing campaign could eventually win the war, the chiefs of the other Allied armed forces thought it could not do so alone, or quickly enough, and that victories at sea and on land would be necessary to defeat Germany and Japan.

The first half of 1942 had seen the military fortunes of Germany and Japan reach their greatest height. Japan conquered most of what became the Greater East Asia Co-Prosperity Sphere, broke all blockades, and controlled most of the world's supply of rubber, tin, and oil. While Japan had advanced, German U-boats traveling in wolf packs decimated Allied shipping and threatened to bring about the collapse of Great Britain, a story told in the next chapter. Through the summer of 1942, German armies won battles against Allied forces in North Africa, Rommel capturing 35,000 British prisoners in a single day, and Hitler's forces in the Soviet Union appeared to be on the verge of causing its collapse. Hitler and his top general were congratulating each other that "the Russian is finished."

A sudden stop to the increasing Japanese hegemony in the Pacific Ocean came in the naval battle of Midway Island, in June 1942, where in just five minutes of concentrated attack, American planes sank 4 large Japanese carriers and the 330 aircraft they had on board. "There was no quicker or more dramatic reversal of power in all history," writes military historian A. J. P. Taylor. "At one moment, the Japanese dominated the Pacific Ocean. Five minutes later they were down to equality in carriers—the essential weapon."

After that the Japanese shelved plans to extend their territory and the war in the Pacific settled into a series of land battles for islands of strategic importance. Part of the reason for the initial American difficulty in retaking these islands was the inaccuracy of airborne bombing; during the Battle of Midway, for instance, American B-17s dropped 377 bombs without hitting a single Japanese ship; and on land the bombing results were not much better, thus permitting Japanese troops to remain in entrenched positions and to force Allied ground troops to root them out, at great cost in lives expended.

One of the bloodiest battles of the war was fought for control of Guadalcanal, in the Solomon Islands. Disease was as much a factor as ordnance, tactics, or supplies. It broke out on the island during the third week of August, and by October there were more than 2,500 cases of malaria per month among American GIs, until more than 90 percent of the Allied forces had been infected. American officers had to stand over the men to make certain they downed their doses of atabrine, which were falsely rumored to foster impotence. The 44,000 Japanese soldiers on the island lacked adequate medical supplies, and of the 28,000 who died there, some 9,000 succumbed to malaria-related causes. In other Pacific Island fighting, captured Japanese hospital records showed that an average of 45 percent of their malaria cases died, as contrasted to less than 1 percent of American military cases; that 55 percent of the Japanese wounded died in their hospitals, and 60 percent of the Japanese soldiers with dysentery and enteritis died, while American frontline hospitals were able to save many more of the wounded and those infected with diarrheal diseases. Shortly, American planes began spraying targeted islands with the new insecticide, DDT, to suppress disease-bearing mosquitoes, at the same time as they dropped bombs to prepare the way for invasion.

Another turning point came in October, in a battle off the Santa Cruz Islands. The carrier *Enterprise* and the battleship *South Dakota* had recently been refitted with Stark Draper's Mark 14 antiaircraft gunsights, and during a fierce Japanese aerial attack the new gunsights helped shoot down more than two dozen Japanese planes, which turned the tide of that battle and gave the American fleet confidence that it could withstand direct massed attacks from the air.

In the fall, in North Africa, the Allies won some victories, seizing territory from German control. Among the prizes were two *Geheimschreiber* machines—German code-making machines for teleprinter transmissions. These were able to encrypt (or decrypt) messages at the rate of twenty-five letters per second and were much more difficult to crack than Enigmas. The British had been after them for more than a year; with typical British deprecation, their product had become known as Tunny for the German Army transmissions, Sturgeon for those of the Luftwaffe—and collectively as "Fish." The captured Fish machines found their way to Bletchley. Attempts were made to crack their codes, then to duplicate and improve the mechanism for generating coded communications.

In November, in retaliation for the Allied offensive in North Africa, Hitler's troops seized the remainder of southern France previously under Vichy control. The Polish mathematician Marian Rejewski and one colleague managed to flee over the Pyrenees into Spain and on to Great Britain. There, rather than being greeted as a great man who had made valuable contributions to the Allies for a decade, he was interned and for some time was not permitted to do further work on the codes he had done so much to break. In Vichy

France the Nazis captured five other former Polish cryptanalysts; their refusal to reveal even under torture that the Allies had broken the Enigma codes was more than courageous—it was heroic; it ensured that the edge provided to the Allies from being able to read the German codes would continue and be enabled to grow in importance and effectiveness.

In Norway, in late 1942, courage could not prevent an Allied failure. In the spring a hydroelectric technician at Norsk Hydro escaped to Great Britain and eleven days later was dropped by parachute back into Norway as a British agent. His reports confirmed earlier Allied suspicions that the Norsk plant was directly connected to the German atomic effort, and the British planned a raid to destroy the plant, whose capacity to produce heavy water the Germans had managed to increase by a factor of ten.

A small group of Norwegian exiles was parachuted into Norway—more than a hundred miles from their target and in terrible weather, although it was only mid-October, and on November 19 the British sent in two gliders filled with dozens of British sappers. One bomber and glider plowed into a mountainside, and the towline of the other glider snapped, causing it to crash. The SS tortured a few survivors until they revealed the mission's objective, and then executed them. The Allies' failed attempt to destroy Norsk Hydro revealed to the Germans the Allies' fear of the German atomic-bomb program and of the heavy-water approach to controlling a chain reaction, which heightened the importance of that approach in the Germans' minds.

The third turning point of 1942 occurred when Soviet troops halted the German advance before Stalingrad. Winter then settled in, bringing death to besiegers as well as to besieged. Eventually troops under the German command, which included those of puppet countries, would lose 850,000 men, 100,000 of them captured, the others killed, and 750,000 Soviet troops would also perish.

To reward Stalin for keeping the Germans occupied in the east and to appease him for the Allies' refusal to invade the Axis-controlled Fortress Europe in 1942 and open a "second front," Churchill offered to share information with the Soviets on what he characterized as all "weapons, devices, or processes which . . . are . . . or in future may be employed . . . for the prosecution of the war against the common enemy." This meant radar, the proximity fuze, jet-engine design, and everything else. American roving ambassador Averell Harriman, working with Churchill, was convinced that the United States ought to agree to this plan, and the U. S. State Department assented—but the American military, and President Roosevelt, did not. Unbeknownst to Harriman or Churchill, Roosevelt had been considering limits to further British access to American atomic research, because of potential British breaches of secrecy, and now he became alarmed that a three-way scientific agreement with the USSR would result in U.S. atomic-research secrets going to the Soviets. And he certainly did not want that happening when American atomic research was on the verge of an important milestone on the road to making a viable bomb: the production of a self-sustaining chain reaction. The knowledge of an impending breakthrough was being very closely held. Even Merle Tuve, a member of the steering committee since its inception, was so discouraged by the ostensible lack of progress in making a bomb that in November 1942 he resigned from the committee as a way of protesting that American resources ought to be directed toward weapons that might win this war, not the next one.

Approaching December 1942 several different methods were under consideration in the United States for producing materials for a bomb, among them a centrifuge method that James B. Conant thought was the "weakest horse" and a uranium "pile" that was supposed to produce plutonium as a by-product of a self-sustaining reaction. The pile was being constructed in a squash court under the stands of the football stadium at the University of Chicago. Conant didn't like the pile method, because he suspected that its proponents—the Italian physicist Enrico Fermi and Arthur Compton, the recently appointed leader of the Chicago team—were more interested in its relevance to a future nuclear power reactor than in making progress toward a bomb. Fermi himself was so uncertain about its safety that he established a "suicide squad" to destroy the atomic pile in the event that something went wrong and the people inside the room were unable to control it. But the pressure to move ahead was considerable, as rumors and intelligence gleanings reported that the Germans had already achieved a self-sustaining nuclear reaction and might well produce a bomb at least a year ahead of any American or British bomb.

On December 2, 1942, however, Fermi and his team erased all doubts by achieving the first self-sustaining nuclear chain reaction in a controlled environment. In the process they also demonstrated the feasibility of using U-238 transmuted into plutonium to make the material for a bomb.

From the Fermi lab in Chicago, Arthur Compton called Conant in Washington and said, "Jim, you'll be interested to know the Italian navigator has just landed in the New World."

"Were the natives friendly?" Conant asked.

"Everyone landed safe and happy," Compton responded.

However, two weeks later many of the scientists and their families began to leave Chicago for shelters many miles away from the uranium-pile experiment—not because they feared that the pile would explode but because there were strong rumors that Hitler was going to launch an attack, possibly with an atomic bomb, on Chicago at Christmastime.

| CHAPTER EIGHT |

Seagoing Science

THE WAR AT sea known as the Battle of the Atlantic was fought in parallel to that on land and with equivalent contributions from science and technology. It was at its height in 1942 and the first half of 1943. Churchill would later characterize it as a "war of groping and drowning, of ambuscade and stratagem, of science and seamanship." In it the Axis would sink 7,486 merchant ships of over 21 million tons, along with 158 British Commonwealth and 29 American Navy warships, and 40,000 Allied seamen. The Allies would sink 781 U-boats, killing 32,000 German submariners. Aside from the immeasurable value of the lives lost, the cost of the sunk merchantmen was staggering. For every two transports and a tanker sunk by a U-boat, more war-related matériel was lost than could be destroyed by three thousand bombers if the same matériel had been warehoused on land.

The sea battle began after Germany joined Japan in declaring war on the United States, when Admiral Dönitz immediately obtained permission to attack U.S. East Coast shipping, with the objective of interrupting supply lines to American factories from South America, the Gulf of Mexico, and the Caribbean. He called his plan Operation Paukenschlag, "drumroll" or "drumbeat."

A de facto state of war had existed between the American and German navies since September 1941, when a U-boat had fired on an American destroyer near Iceland and President Roosevelt had declared that U.S. ships would thereafter make an "active defense" against such "legal and moral piracy" while patrolling the Denmark Strait between Iceland and Greenland. During October, U-boats had sunk several American ships in that area, among them the *Reuben James*, killing 160 of its crew.

Hitler had once said, "On land I am a hero, but at sea I am a coward." That may have been why he never fully understood or paid full attention to the war at sea. Although the overall size of the U-boat fleet had doubled since September 1939, there were still not very many U-boats, and in terms of scientific-warfare implementation—torpedoes, detection devices, decoys, deck arms, communications—the submarines in service were not much better equipped than those at the outset of the war were. They did carry a few gadgets: weighted buoy decoys that could be released to confuse the enemy's sonar by emitting chemically produced bubbles, and sonar-deflecting wire-mesh grids fitted onto the sides of the U-boats.

Toward the end of January 1942, when Dönitz's first five U-boats arrived off Washington, Baltimore, Philadelphia, and New York, the commanders sent home signals of astonishment at the rich and easy pickings that awaited—"Enough to keep ten or twenty U-boats busy," a U-boat commander wrote in his diary; during just one night off Cape Hatteras, near Washington, he was able to sink three large steamers and hit a fourth. Most coastal ships steamed slowly along well-marked lanes up and down the seaboard, their lights on (and channel buoy lights on) even at night. They were even more vulnerable because they had no escort vessels, did not travel in convoys for protection, and were starkly silhouetted at night by cities that continued to keep their lights on after dark. The German subs proceeded to massacre East Coast shipping. People on beaches, unaided by binoculars, could watch U-boats hit merchantmen with torpedoes; oil slicks and occasional dead bodies would float ashore. To conserve torpedoes the U-boats would surface and rake defenseless tankers and cargo ships with deck guns. Between January and April of 1942, U-boats sank nearly a hundred merchantmen off the East Coast, and although American factories were not shuttered by the loss of raw materials those sinkings caused, production was seriously affected.

German submariners called this the "happy time," because they operated along the East Coast almost without opposition. The American Navy's attention was then focused on the Pacific, where Admiral Ernest J. King, recently appointed commander in chief of naval forces, was trying to stop the Japanese naval advance.

King had been on the verge of retirement when the war in Europe began, and he had not expected to be called to action. A hard drinker and womanizer, enjoying what his biographer described as "a private life of notable gaudiness," he swore off alcohol after learning of the fate of the *Reuben James.* He was characterized by his daughter as "the most even-tempered man in the Navy—he is always in a rage." In March 1942 General Dwight Eisenhower wrote in his diary, "One thing that might help win this war is to get someone to shoot King," who was refusing to shortchange the Pacific fleet for a Germany-first strategy and was also at odds with General Marshall over a unified command in the Pacific theater.

In the person of Admiral Ernest King there was everything that scientists—and, indeed, political leaders—disliked about military men, but there was also intelligence and the capacity to change. Eisenhower was wrong to complain about King's preference for a two-ocean war, for it was King's insistence on early pressure on Japan that prevented that country from consolidating its Greater East Asia Co-Prosperity Sphere and from allocating the resources necessary to develop superior weapons before the war's end.

Nevertheless, in the opening months of American involvement in the war, King made some cardinal errors in regard to the U-boats off the East Cost, the most important being the delegation of the task of countering those submarines to a Naval Academy classmate, whose antiquated antisubmarine warfare forces were scattered from Bar Harbor to Key West and were unable to kill a single U-boat during the first several months. Also during that time American shipping tried to cross the Atlantic piecemeal, as King did not see fit to institute mandatory convoying of goods and armaments to the British Isles. Some 118 merchantmen traveling independently were sunk by U-boats, while only 20 traveling under escort were sunk—a statistic that troubled President Roosevelt, who needled King about it.

King's initial refusal to institute convoy procedures for cross-Atlantic and coastal shipping, writes German naval historian Juergen Rohwer, "was without doubt one of the greatest mistakes in the Allied conduct of the Battle of the Atlantic." It also stirred the confrontation shaping up between King and Vannevar Bush, a clash between scientific and conventional military thinking. Its most vitriolic stage would come later on, but in 1942 there was a first skirmish, over Bush's recommendation to King that research should be undertaken on countermeasures to acoustic torpedoes. The Germans had not yet used such torpedoes, but civilian scientists predicted they would; King grudgingly gave his approval to the research, so long as Bush's agency funded it. Bush did, and he also commissioned basic research on other ASW matters: sound transmission, long-range magnetic detection, fluid dynamics, alternate propulsion systems, and explosives.

In the late spring of 1942, after several months of uncontrollable losses of transatlantic shipping, Admiral King woke up and started cross-Atlantic and coastal convoys. Convoying reduced the losses, even though the only escorts available were older ships with too-wide turning radiuses that lacked the most advanced weaponry and detection systems.

In reaction to the use of convoys, Dönitz simply moved his coastal operations southward, into the Caribbean and the Gulf of Mexico, where the U-boats' devastating success resumed. When Allied ASW forces started hunting individual submarines, Dönitz shifted to the "wolf pack" technique, in which three or more U-boats would act in concert, trailing a convoy and picking off stragglers on the outskirts of the group.

A secret of the U-boats' success was that they had ten to twenty hours' prior knowledge of Allied convoy movements: German Intelligence had broken British Naval Cipher No. 3, then jointly used by British, Canadian, and American convoy and sub-hunting forces. The Allies would not realize that their code had been compromised for another eighteen months. Nor were the Allies able to steal similar information on German plans after February 1, 1942, when a fourth wheel was added to Enigma coding machines used on-board the U-boats.

By February the ongoing struggle between Raeder and Dönitz for command of resources—and of the Germany Navy itself—had become pronounced, the Grand Admiral's fortunes declining in Hitler's eyes as a result of the inability of his surface ships to counter the British surface navy, while the sub-fleet commander's were rising because of the "happy time."

Hitler ordered an increase in the manufacture of U-boats, and Speer saw to it that this time the order was carried out. Soon two dozen new U-boats were launched each month. However, the increase was only in number of subs, not in terms of their armament, endurance, underwater speed, or length of time they could remain submerged. Since the mid-1930s Germany had possessed plans for a turbine-powered propulsion system that could double underwater traveling speed to 26 knots, faster than that of convoys, but without Raeder's support the turbine design had languished. In 1942 Dönitz was finally apprised of the turbine design, saw a prototype, and urged its production, but turbine-powered subs were not expected at sea until 1945. Germany similarly lacked a long-range bomber—as we have seen, the decision not to build one dated to 1937—and so in 1942 had no bombers able to locate convoys or to provide cover for subs as they traveled through the Bay of Biscay or made dangerously exposed refueling operations in the mid-Atlantic.

The Allies faced fundamental problems in the Battle for the Atlantic to which science could contribute some answers. The most pressing problem was a gap of hundreds of sea miles in the North Atlantic, where the wolf packs could do their hunting free of effective harassment by Allied antisub forces. The gap lay between the area of sea lanes that could be covered by planes based in the British Isles and those that could be covered by aircraft based on the outskirts of North America. To patrol it the British had only five "very long range" planes, called Liberator 1. Moreover, the British destroyers that could have acted as convoy escorts for transatlantic crossings in 1942 were allocated to the Mediterranean for use in the fighting in North Africa, while the American destroyers were being used in the Pacific Ocean.

Another basic problem was the limited ability of Allied ASW forces to sink submarines once U-boats had been spotted. Communication and coordination between ships and aircraft was abysmal. Depth charges dropped by ships were only partially effective. Attacks from the air produced damage mostly if the attacked submarines were on the surface; usually subs were able to note the approach of an aircraft by radar and had adequate time to submerge before a plane came close enough to shoot at them. As Dönitz bragged to his commanders, "the U-boat has no more to fear from aircraft than a mole from a crow."

In Great Britain the task of devising new ways to counter the U-boats fell principally to Operations Research. Many of the best notions emerged from the "Sunday soviets," freewheeling discussions among all levels of scientists and military experts at the Telecommunications Research Establishment, which were brought into being and attended by Blackett and other leftist scientists, based on the ideas they had absorbed from Soviet scientists in the interwar years.

An interesting mix had jelled in Great Britain, amalgamating the intellectual power of the scientists on the left, like Blackett, Bernal, and Cockcroft—who were spurred to action by their belief in the "science in the service of society" ideas of the Soviet system—with those on the right, like Lindemann, Tizard, A. V. Hill, and Reg Jones—who were equally energized to action by the belief that only a strong military fueled by scientific advances could overcome the Nazi war machine.

The conjunction of the two assisted the OR analysts in getting their views accepted and put into practice by the military when, as frequently happened, the scientists' analysis led them to recommendations contrary to the accepted wisdom that was the basis of RAF and Royal Navy antisubmarine-warfare tactics. For instance, their analysis of data showed that planes flying low had a better chance of spotting surfaced submarines (and converting those spottings into sinkings) than did those flying high; that refueling during flight—though unpopular within the RAF—could produce better coverage than having planes return to bases to be refueled; and that traditional ways of aiming torpedoes and depth charges at submarines were not doing the job properly. Blackett, who had already successfully reduced the amount of ammunition required to bring down a plane from 20,000 rounds to 4,000, was someone whom the air corps establishment had reason to respect and listen to, and so these recommendations were heeded.

Analysis showed that the key to more sub kills by planes was more time in the air and greater efficiency of flying. Many benefits, the number crunching suggested, would accrue to airplanes meeting convoys at sea and patrolling on their behalf, so that the grouped ships could alter course to avoid wolf packs or steam in tighter formation and make themselves into more difficult targets. But because airborne navigation until then had been done by "dead reckoning," not by more accurate electronic means, the "not-met" rate for planes finding convoys was 8.5 percent for each 100 miles, which meant that at 600 miles from base—where the danger from subs was greatest—convoys were being missed as often as they were met, drastically cutting the patrols' effectiveness. Using mathematical estimation and prediction tables, the British reduced the not-met rate by half and lowered it further when new radar navigation sets became available. To prevent German subs from using intercepted signals to locate convoys, OR suggested that ships break radio silence only for short transmission bursts and send information not about their own positions but about those of the planes, which could then be radioed back to them.

A study of the reasons for canceling air patrols yielded the understanding that the most limiting factor was not bad weather but the amount of maintenance and spare parts available; scheduling maintenance for bad-weather days and having more spare parts in inventory increased patrol time in the air and raised the number of subs successfully hunted. An analysis of 7,000 plane "contacts" with submarines revealed a counterintuitive finding: that flying higher did not necessarily result in greater range covered, submarines spotted, or subs sunk. The reasons had to do with visual acuity at different heights. The study led to a directive that planes should fly below 2,000 feet, and even lower when seas were choppy. A similar directive was issued to radar-set operators: no more than half an hour at a sitting in front of a set without a break, because after half an hour fatigue compromised the watcher's efficiency.

The Royal Navy had based its prior practice on the facts that U-boats could not stay submerged for very long or move very far while submerged. The OR scientists eventually convinced the admirals that it was a waste of time and matériel to attack subs that had been submerged for more than a minute or two. They also changed the pattern of laying down depth charges. In a normal deployment, depth charges were catapulted off the rear of a moving ship that was trying to stay ahead of a diving submarine. Using the data from Operations Research analysis to suggest the best way, time, and place to hurt a sub, Blackett and others designed two new weapons—the 100-pound depth-charge bomb, with shaped explosives, to be dropped from an airplane, and the ship-based "Hedgehog."

The first trial of the airborne bomb resulted in an explosion that blew up the plane and killed the pilot; Blackett abandoned the effort, and it fell to Edward Terrell, the barrister who had been working for the Navy on armor protection, and to Willis Jefferis, the oddball inventor favored by Cherwell. Terrell tried 35-pound bombs, in both the airborne and ship-borne configurations, marrying them to the hollow, shaped-charge projectiles that had been used before the war for mining. The hollow charges had explosives placed behind a cone of steel, so that when fired, the heat and directionality of the explosive propelled the cone as though it were a jet of metal traveling at 7,000 feet per second. This power was enough to penetrate both the outer hull of a submarine and the ballast water and oil tanks that protected the sub's inner hull.

The Hedgehog could throw two dozen of these shaped charges in a circular pattern a hundred yards wide and was designed to do so in front of a moving ship rather than behind it. The charges were also set to explode on contact rather than at a specific depth. Armed with the newly developed Canadian explosive RDX,[1] the Hedgehogs had a much higher kill ratio than ordinary depth charges. Their efficiency was further improved by an OR directive to discount a sub's forward motion when aiming a cluster from the Hedgehogs and to target only the conning tower.

American mathematicians and other scientists contributed to heightening what ASW forces could accomplish in several ways, for instance by making a "Mousetrap," a multiple-rocket device that, like the Hedgehog, fired its bombs ahead of the small sub-chaser boats in which it was seated. Louis Slichter, a professor of geophysics at MIT and Caltech who designed the Mousetrap, also invented a "retro" rocket that could be fired—backward—from a plane that spotted a sub directly beneath it, in a way that directed the missile to that spot and compensated for the plane's forward motion. The American Navy's Operations Research group, headed by MIT physicist Philip Morse, designed better flight-search patterns and plane-ship coordination methods. Shore-based planes were used as much as possible to cover wide search areas, which permitted ship-based planes to remain closer to convoys, where they had the best chance of finding and sinking submarines. OR analysis on both sides of the Atlantic also determined that the best moment for sub-hunters to successfully attack U-boats was when the U-boats were attacking a convoy and had to stay in position in order to fire their torpedoes.

The Allies also employed more technological ASW tools, such as powerful searchlights mounted under planes. U-boats seeking to reach their hunting grounds in the open Atlantic and to conserve power, fuel, and maximize speed had been crossing the Bay of Biscay

[1]It had taken Canadian, American, and British scientists more than a year to convince the military establishments to use the new explosive, though RDX was 40 percent more powerful than TNT. By July 1943 a Kingsport, Tennessee, plant was manufacturing 170 tons of it per day. To prevent the enemy from learning about the new explosive, shipments of it were labeled as golf balls; also, information about RDX was deliberately not shared with the USSR.

at night and on the surface. Allied searchlights (called "Leigh lights" after their inventor) illuminated targets for sub-hunting planes' torpedoes and depth charges; with this equipment the Allies sank dozens of subs in the Bay of Biscay and drove most of the remaining U-boats under the water for the duration of their crossings, while forcing others to chance making the surface transit in daylight, when U-boats had a greater likelihood of spotting and shooting down Allied sub-hunting planes before the planes could spot them.

The RAF, fearful of German ship-based searchlights, had painted the undersides of their own planes black; after an OR study the British decreed that the planes' undersides should be white, to protect against the statistically larger danger of the planes' being detected in daylight against a backdrop of white clouds. In the Mediterranean the underbelly color was to be "dark sea gray," and in the Atlantic near the United States shiny black, which at night reflected lights in a way that kept the planes camouflaged.

The OR experts had long expected the Germans to outfit their ships with receivers able to detect the transmission of Allied radar, but that did not happen until the autumn of 1942, when the Germans installed the Metax system. Metax contributed to German U-boats sinking more ships per month (over half a million tons) than could be built by Allied shipyards, while German shipyards completed more U-boats per month than Allied ASW could sink. Hitler considered acceptable an "exchange rate" in which for every 50,000 tons of Allied merchantmen sunk, the Allies sank one U-boat.

Prior to the outset of war, Great Britain imported 60 million tons of food and other supplies annually, excluding oil; in 1942–43 the total had dropped to under 20 million tons, close to the minimum safety level for sustaining the population and its war effort. Cherwell's multipronged approach to this problem brought scientific thought to bear on such matters as condensing the use of shipping space to transport supplies per ship—for instance, shipping trucks in CKD or "completely knocked-down" (unassembled) condition, buying more ships from American shipyards, and cutting down on supplies sent to garrisons abroad, particularly in the Middle East. These efforts produced salutary results but were also blamed for exacerbating an enormous famine in India in 1943, in which 3 million people died.[2]

The need of the British to prevent the U-boats from cutting the supply lines that were keeping Great Britain alive was so desperate that it midwifed the birth of a "scientific" project to bridge the thousand-mile gap in the Atlantic. A gigantic floating airfield made of ice was proposed by a crackpot inventor, failed spy, Cambridge dropout and left-wing sympathizer named Geoffrey Pyke. He managed to convince first Lord Louis Mountbatten (a cousin of King George VI) and then Churchill that this idea was brilliant and feasible. Mountbatten had compiled a dreadful war record—his arrogance had resulted in three ships sunk underneath him, and he bore a major responsibility for the disastrous Canadian raid at Dieppe in which 3,367 of 4,000 soldiers were killed, wounded, or captured—but the popular press perceived him as a hero, and Churchill made him head of Combined Operations.

Mountbatten had been charmed by Pyke's earlier ideas, such as to blow compressed air under the German battleship *Tirpitz* and make it sink by causing it to fall into a bubble, and to develop a machine that could travel over snow by means of a screw mechanism that would bore through the snow. Churchill liked the latter one as well, and he dispatched Pyke to the United States to discuss manufacture of the vehicle—one of Churchill's usual tactics when confronted with a new and obviously expensive idea. While in New York, Pyke conferred with a Brooklyn Polytech professor who had determined that

[2]Controversy still rages over whether it was Cherwell/Churchill policies, ineptness of administration in India, or mistakes made by the United States in allocating ships and cargo, that exacerbated the shortages of rice and other foods; the famine was most likely due to all of these factors.

ice mixed with a few percent of sawdust became tough and very slow to melt—and that gave Pyke the idea of calling the mixture "Pykrete" and using it to build floating airfields called "Habakkuks," as well as freighters and assault ships. His fevered brain imagined a fleet of Pykrete ships sailing into German, Italian, or Japanese home ports and spraying the enemy vessels with supercooled water to freeze them into immobility.

This science-fictional notion, written up in a 232-page memo to Mountbatten, was passed along to two other Combined Operations advisers, J. D. Bernal and Solly Zuckerman. Though Zuckerman dismissed the scheme as lunacy, Bernal recommended it. Since there seems no good reason that any competent scientist would back this project—and Bernal was competent—historians have speculated why he would have agreed to Pyke's idea; one answer is that both he and Pyke were Marxists, and Bernal owed him some sort of comradely loyalty; another is that Bernal was simply tweaking the establishment with an eye toward eventually embarrassing both Mountbatten and Churchill.

Since there was a palpable need for such a device as an airfield on which sub-hunting planes could refuel in the mid-Atlantic, Mountbatten seized on the idea and wrote to Churchill about the potential of a Habakkuk "immune to bombs, mines and torpedoes," and Churchill sent a memo to the Chiefs of Staff enjoining them to assist Mountbatten in making a Habakkuk, not of Pykrete but of natural ice. "Go to an ice field in the north which is six or seven feet thick," Churchill instructed his service chiefs: "Cut out the pattern of the [ice] ship on the surface; bring the right number of pumping appliances to the different sides of the ice-deck, spray salt water on continuously so as to increase the thickness. . . ."

A hundred-page memo by J. D. Bernal on the subject was circulated at a meeting of Canadian scientists, who were incredulous. Didn't anyone realize that the sides of this ice ship would have to be 50 feet high in order to prevent it from being swamped? And that 50-

foot-high sides could be constructed only from a 500-foot-thick sheet of ice? In Great Britain, Cherwell and Sir Arthur Goodeve, who was in charge of the Admiralty's scientific research, had to sit on a Habakkuk committee and repeatedly ridicule the notion—but could not entirely kill it. When using natural ice was shown to be not feasible, Mountbatten switched to the notion of using Pykrete, a cake of which he demonstrated to Churchill when the prime minister was in his bath at Chequers. Churchill sent Bernal and Pyke to Canada to supervise the construction of a Pykrete model on Lake Patricia, in Saskatchewan. Parts of the demonstration were impressive: bullets fired (by Bernal) at the Pykrete-based beams bounced off with no effect. But an upgrade from model to actual floating airfield was impossible: The airfield would have had to weigh 2.2 million tons and be ten city blocks long—the largest extant ship was three blocks long—and it would have absorbed the entire output of the Canadian wood-pulp industry for a year, leaving none for use in newspapers. Canada officially refused to complete the project and threw it back into Churchill's lap, where it languished—for a while.

The fortunes of the Battle of the Atlantic began to change, very slowly, in the late months of 1942, when—after three years of war—the Allies' most important secret detection device was introduced on a limited number of British ASW ships and planes: microwave radar. With its much shorter bandwidths, microwave radar was able to locate surfaced submarines at greater distances and to do so without revealing its own presence, since German receivers did not recognize its bandwidths. The new radar gathered information that permitted convoys to change routes and avoid wolf packs. As a result of microwave radar, sinkings by the 40 U-boats in the waters of the North Atlantic in November 1942 dropped to a low of 23 ships, although the overall total for the month was a staggering 106 ships. Shortly thereafter the tonnage lost to U-boats in the North Atlantic rose again as the Allies' attention was diverted to North Africa and as more and more U-boats were deployed.

Another factor was the introduction of submarine tanker ships called "milch cows," which carried fuel, spare torpedoes, and other supplies to enable U-boats to stay at sea longer and to range farther from their land bases. The milch cows were a good idea, but fuzzy German thinking about scientific matters countered this new innovation and added to the U-boats' difficulties in battle. Standing orders to U-boats were to report back to headquarters every time one spotted a convoy or an individual target; to prevent the Allies from using these signals to locate the subs, the Germans sent transmissions in short bursts of under thirty seconds—a technique based on the mistaken assumption that the Allies could not have developed a radar able to pinpoint a location from which short-burst signals were sent. But the Allies did have such a new High Frequency Direction Finder (HFDF or "huff-duff") radar. Several Allied ships that were miles apart could join together in using their HFDF to identify a transmission source, and the triangulation provided a very precise location of an enemy submarine. The Germans photographed HFDF antennae aboard Allied ships but were never able to identify their function or to connect the antennae with the U-boats' difficulty in hiding from ASW hunters.

A second Allied technique unknown to the Germans permitted identification of individual submarines by the characteristics of the signal being sent; "radio fingerprinting" identified each transmitter by using high-speed photography of lines on cathode-ray tubes.

A third technique involved the mass accumulation of data and coordinated attempts at decryption. A set of stations in North Africa, Iceland, and Bermuda, as well as up and down the North American coastline intercepted sub transmissions and sent the data over secure phone lines to London, Ottawa, and Washington, where teams of cryptanalysts worked on them. After the German code was broken once more in December 1942, the submarines' radio traffic yielded great quantities of useful information about the U-boats' positions, their physical condition, and their amount of fuel supply.

An OR study suggested that each German U-boat was already being subjected to Allied attack four times a year and that a 10 percent increase in the "kill rate" of ASW would be enough to cause Hitler to lose more submarines per month than he could build.

How to raise that kill rate? By improving every aspect of making war at sea. The plans for the Oerlikon 20-millimeter antiaircraft gun were smuggled out of Switzerland to Great Britain, and manufacture of these guns and of the Swedish Bofors 40-millimeter guns raised the accuracy and range of armaments on Allied ships. Radiotelephony between ships was improved, as were procedures for damage control—Allied ships, particularly American ones, became more able to resist sinking and to return to service after being damaged than those of Germany, Italy, or Japan. Other Allied scientific and technical innovations include gun liners that extended the lives of gun barrels and the "frangible" bullet, capable of being used by many different weapons, which freed factories from the need to manufacture special bullets for each weapon in the arsenal.

Allied torpedoes also needed improving. American torpedoes were then powered by steam, were easily seen because of their white wakes, and had magnetic detonators that seldom worked; Japanese ships regularly returned to port with unexploded torpedoes stuck in their hulls. The Japanese "long-lance" torpedo, propelled by a liquid oxygen mixture, had a speed of 49 knots and a range of 11 miles, which meant it could be launched from well outside the reach of any surface ship's armament. It was several times as deadly as Allied torpedoes—or, for that matter, German ones. National pride prevented the Germans from adopting the Japanese long-lance, but their own electric-powered torpedoes were effective against Allied shipping.

An unexploded German electric torpedo picked up on a New Jersey beach in mid-1942 became an object of curiosity as well as of research, for it revealed a crucial distinction between the American and German ways of science and technology. The German torpedo was exquisitely designed but almost handmade; it could not be modified or repaired because it lacked uniform, machine-made parts. American engineers were very quickly able to understand its workings and to duplicate the basic design, but they did not stop there; they then altered the torpedo design so that it could operate more easily—for instance, the American version's battery was activated by salt water—and so that the components of the torpedo could be stamped out in quantity. The result was not the best or the most exquisitely made torpedo in the world, but it was one that could be readily and rapidly manufactured in the United States.

Within a short period of time, American torpedoes from this modified German design were used in the Atlantic, but they found their best use in the Pacific, where they were able to decimate Japanese shipping. The task of the American torpedo in the Pacific was made easier by the refusal of the owners of Japanese merchant ships to allow them to travel in convoys—because the lower speed of a convoy cut into the profit potential of a voyage. The massacre of Japanese merchantmen by American submarines did what the German U-boats never fully succeeded in doing—crippled a major belligerent's war-making capacity.

In early 1943 the Germans introduced the T-4 acoustic homing torpedo, an electric-powered torpedo slowed to half speed to allow it to home in on its target. It did not work very well, and Vannevar Bush's perspicacity in having set researchers to the task of countering the possibility of such a torpedo now paid off. American researchers came up with free-floating decoys that emitted sounds similar to those made by ships, decoys that could attract the new torpedoes and make them explode where they could do no harm.

In January 1943, after the failure of German surface ships to win a battle with British ships convoying merchantmen toward Murmansk,

Hitler called Admiral Raeder on the carpet and berated him with figures and myths that blamed the German Navy for all the Reich's troubles. Raeder "considered it beneath my dignity to challenge the details of this completely fabricated story" and resigned, effective January 30, the tenth anniversary of the founding of the Third Reich.

Hitler chose Dönitz as Raeder's successor, and Dönitz used the opportunity of his promotion to switch the Navy's emphasis from surface ships to submarines. A directive of February 2, 1943, announced that construction of battleships, heavy cruisers, aircraft carriers, and troop transports would cease, that many ships would be "paid off"—taken out of service permanently and used for scrap, their guns to be dismounted and set up for coastal defense. Simultaneously he raised production targets for U-boats, reinforced their hulls, provided them with snorkels to lengthen the time they could stay under the waves,[3] attempted to increase their technological sophistication, and haggled with Göring for more air cover so that U-boats could travel safely toward their hunting area. The tonnage sunk by U-boats stayed low through January 1943 in bad weather in the North Atlantic.

At Casablanca in January 1943, Roosevelt, Churchill, and Stalin agreed that countering the U-boats was the most urgent priority. At the suggestion of Admiral King, responsibility for convoying was carved up so that American, Canadian, and British ships all had sectors of the ocean to patrol. These administrative maneuvers helped, but as the losses to U-boats again began to soar—in February the U-boats sank 108 Allied ships and in the first twenty days of March sank another 107—it became clear that better convoying alone would not win the Battle of the Atlantic. An alarmed Churchill told Parliament, then agitating for an invasion of Europe, "The defeat of the U-boat is the prelude to all offensive operations." "There was no doubt in my mind that we were headed for catastrophe," Vannevar

[3]U-boats without snorkels had to surface at least four hours out of every twenty-four to recharge their batteries; snorkels permitted recharging while the submarine remained underwater.

Bush later wrote about February–March 1943. "If U-boat success continued to climb, England could be starved out, the U.S. could mount no overseas attack on the Nazi power, Russia could certainly not resist alone."

During March the Allies sank 14 U-boats, but 27 new ones were commissioned, bringing the total in service to over 400, an average of 116 a day in the Atlantic, nearly 50 of these on station. In January each U-boat had sunk an average of 129 tons of shipping; by March the figure reached 230 tons per U-boat.

Great Britain rushed into wider use the new microwave radars and Hedgehogs, and hurried the delivery of new Liberator very-long-range planes to bridge the thousand-mile gap in the North Atlantic. But American ships and planes on ASW patrol did not immediately adopt all the new weaponry, and because the Americans had responsibility for the convoys most of the way across the Atlantic, the peril from the U-boats continued to deepen.

American ASW was hampered by Admiral King's refusal to accept the new scientific devices for hunting subs. "Escort is not just one way of handling the submarine menace," King wrote, "it is the only way that gives any promise of success. The so-called hunting and patrol operations have time and again proved futile." King would not install the new microwave radar on sub hunters nor agree to use Hedgehogs. Vannevar Bush became fearful that the advantage that microwave radar gave to the Allies would be dissipated as soon as the Nazis learned about the new bandwidth and devised a receiver to detect its presence. He enlisted the Army to do what the Navy would not. With the assistance of Hap Arnold and of Edward Bowles of MIT, Rad Lab–developed search radar was installed on ten B-18s flying out of Langley Field on coastal patrol. On their first night of patrol, the Army air group spotted three submarines and sank one of them. Arnold immediately put Bowles in charge of all communications, radar, and electronics for the Army Air Force. But the Navy still would not mount search radar or advanced antisub ordnance on its own planes or ships, and Bush was unable to move King from intransigence—not even when he and Bowles convinced Secretary of War Stimson to take up the cause with President Roosevelt.[4]

So in late March 1943, when at a private lunch FDR asked Bush about the submarine menace, Bush pulled no punches, telling the president that King was the bottleneck; then Bush went back to his office and told King's staff what he had done. King was apoplectic, feeling that Bush was trying to take over and remake naval strategy. Bush then wrote King a six-page letter that painted a logical and clear picture of the situation. "We are in combat with a resourceful and technically competent enemy, and are engaged in a race between techniques," Bush wrote. He pointed out that the new Allied devices meant a complete reorientation of traditional ASW, which should be done by experienced naval officers in concert with the people who understood the science of the devices. That letter and a follow-up meeting of the two men provided King with a way to cooperate with Bush and his scientists without diminishing King's own authority or his ability to conduct naval operations, and led to one of the most innovative managerial developments of the war, King's creation of the Navy's Tenth Fleet. It was a fleet that had no ships of its own but consolidated all ASW activities in the Atlantic under a single authority, one that soon included Morse's Operations Research group and permitted scientists to be aboard ships and planes that were in the process of hunting subs. The Tenth Fleet was the beginning of extensive cooperation between the Navy and civilian scientists and of bringing Operations Research to bear on the full panoply of tasks that the military had to accomplish.

On April 1, 1943, the situation in the Atlantic was grim for the Allies and glorious for the Germans. But in the next two months, and

[4]Stimson had taken a ride in a microwave-radar-equipped bomber himself, in July 1942, and had watched it accurately locate a distant ship. He went back to his desk and wrote notes to Generals George C. Marshall and Hap Arnold: "I've seen the new radar equipment. Why haven't you?"

very suddenly, four breakthroughs came together, and their conjunction reversed the fortunes of the two sides. The four were the decoding of German radio traffic to and from the U-boats, microwave and HFDF radar, ASW patrol tactics (and the number of available planes), and new weaponry such as the Hedgehogs and the Mark 24 air-dropped torpedo mine.

The code breakthrough by the Allies of the German radio traffic and the ability of the Allies not only to read the messages but to distribute the decrypted and translated information to ships and planes at sea, usually within twenty-four hours, was a great boon to the convoys in terms of their taking evasive action, and to the sub-hunters, which were able to know in advance such vital information as the number, location, and state of readiness of the subs they were facing. Germany was also able to decrypt the Allies' codes, which the Allies did not know, but in the game of cryptanalytic chess played over and under the Atlantic, the Allied code break proved to be a greater advantage than that of the Germans.

In later years Dönitz would attribute the defeat of the U-boats in the Battle of the Atlantic solely to airpower and microwave radar, never acknowledging that Allied code breaking also had played a role. On the other hand, later historians would attribute most of the Allies' success to Ultra and the code breakers. But W. J. R. Gardner, the author of a recent reevaluation of Ultra's role in the war, suggests that in regard to the Battle of the Atlantic, Ultra has been overrated, since the code breaks made little difference until the Allies could deploy adequate numbers of microwave radar sets, HFDF, radiophones for communication and coordination among ASW units, and destroyers to protect the convoys.

As important to the Allied victory, the Germans could not detect Allied short bandwidth microwave or HFDF radar, did not understand the new sub-hunting patterns that these radars and the Hedgehogs permitted, and never believed that the Allies had broken the German naval codes.

Microwave radar and HFDF combined to deter U-boat attacks even when no air cover was present, since destroyers were sent along the bearing of the transmission, and their presence usually forced the U-boat to submerge and take evasive actions rather than to aim torpedoes at merchantmen. Closer to the British coast, British sub-hunting planes employed the new H_2S locating radar and with it managed to sink a substantial number of U-boats.

The new offensive weaponry contributed as well. The Hedgehog was a British production from start to end, but the new Mark 24 torpedo was an American production.

Research on undersea warfare had been going on in the United States at locations as varied as the University of Washington, the Hemenway Gymnasium near the Harvard campus, Bell Labs, and on the sixty-fourth floor of the Empire State Building in New York City. The standard of the American torpedo fleet was the Mark 18, the electric fish adapted from the captured German model, but recently work had been completed on a torpedo to be dropped from an airplane like a depth charge.

After rejecting control of a torpedo based on radio, magnetism, heat-seeking or pressure-sensitive technologies, the American researchers agreed that acoustics offered the best possibility for a homing torpedo and began to design the Mark 24. General Electric made the motor—adapted from that of a commercial washing machine—but there were dozens more technological problems to solve, such as how the device would "acquire" submarine sounds, whether it could be designed to drop through the water faster than five feet per second, and how best to steer it. Decisions about such matters were put into the hands of young researchers who could hardly believe their good fortune in being able to perform research that would clearly make a difference in the war; had it been a time of peace, they came to believe, they would have had less interesting jobs and would have been forced to wait patiently to accumulate enough seniority to make important decisions. Four magneto-based hydrophones allowed the Mark 24 to hunt submarines, a much harder task than hunting surface ships.

When all the problems had been solved, production models of the Mark 24 were sent into battle, but carefully. It was so new and secret that orders were to drop it on only one submarine at a time, so that an example could not be captured, taken to Germany, and made to reveal its secrets. Within days one Mark 24 destroyed a U-boat off the American coast, another a U-boat in the North Atlantic, and a third a U-boat south of the Azores. In later years a historian of torpedoes would credit the Mark 24 as "the thing that broke the back of the German submarine program."[5]

The entire experience of the Allies in the Battle of the Atlantic showed that science-based weaponry or technologies were no more than laboratory toys unless and until they were put to use by the military and manufactured in sufficient quantities to make a difference in battle. For instance, American laboratories had devised an airborne fire director for torpedoes launched from planes; it worked well in tests, but the scientists could never convince ASW plane pilots or bombardiers that it was any better than their experienced eyes—and it was never used in battle during World War II. Microwave radar was used, and used well. The bypassing of production bottlenecks for the microwave sets, the sets' enthusiastic adoption by the Navy after a slow start, the adapting of the technology for use in airplanes as well as on ships, and wide distribution of microwave sets throughout the Allied forces helped transform a laboratory breakthrough into a winning weapon.

As for air-patrol tactics, the key to their full use in the spring of 1943 was the beefing up of the fleet of Liberator planes from 11 to 80, so that they could provide much more coverage of convoys from Iceland, Greenland, and Ireland. The flight patterns and maintenance schedules suggested by Operations Research, the coordination with surface vessels that Bush and others had been trying to

[5]A similar claim was made for H_2S.

achieve, and the enhanced equipment of the planes raised the level of the Allied attacks on the U-boats. A year earlier OR had opined that if the Allies could raise by 10 percent the effectiveness of attacks on the U-boats, that would be enough to overcome the rate at which German shipyards were completing new ones. Soon the Allies' kill rate began to rise, and the question became whether the Allies' effectiveness in killing U-boats would escalate fast enough to prevent disaster.

A turning point was reached in the last week in April and the first week in May with the westward-bound convoy named ONS-5, which engaged German submarines in the largest convoy battle of the war. Allied convoys had been growing steadily in size since Operations Research had shown that what best aided the ability of the convoy to get through to its destination without losses was the ratio of escort vessels to merchantmen rather than the absolute size of the convoy, which brought about bigger and better-escorted convoys. ONS-5 consisted of more than fifty vessels.

As the battle of ONS-5 began, the Allies lost their cryptographic edge and were without it for the first week, though Intelligence still estimated sixty U-boats in the North Atlantic waters. By the time the second week was over, forty U-boats had attempted at one time or another to attack the convoy, and they had sunk thirteen merchant ships. But the Allies had sunk six of the U-boats, had damaged five others severely, and had inflicted lesser damage on a further dozen, a trade-off that Allied Intelligence deemed "heavy punishment for results which in the circumstances must have seemed meager" to the Germany Navy. Reports to Dönitz stressed the depredations of Allied radar, which was "robbing the submarine of her most important characteristic—ability to remain undetected," and describing how long-range Allied aircraft "always forced our submarines to lag hopelessly behind the convoy and prevented them from achieving any successes." This observation was a slap in the face to Dönitz, who had so blithely boasted that the subs had less to fear from aircraft than "a mole from a crow."

Now German imagination on the subject of Allied science went wild, producing a report stating that the Allies had some sort of device—perhaps an infrared sensor, perhaps an enhanced radar—that allowed a single Allied plane to understand the location of an entire 600-mile-long U-boat patrol line. Against such a mythic superdetector the Germans could mount no defense. And no offense either: the next westward Allied convoy, ONS-6, was attacked by U-boats, but there were no sinkings; in mid-May, despite the Germans having very good intelligence about the locations, speeds, and courses of two Allied convoys, the U-boats were forced after repeated attempts to break off operations.

Of greater importance to the Allies, analysis of German communications in those aborted attacks showed Intelligence that Allied codes had been broken, a fact they had not previously understood.

The introduction of new Allied codes, which diminished German intelligence about the locations of convoys, may have been the last straw for Dönitz. Between April 10 and May 23, 1943, his submarines had fought a dozen extensive convoy battles and had sunk only twenty-nine ships while losing twenty-two U-boats, including one in which Admiral Dönitz's son served. Nine other U-boats had also been sunk in May, for a total of thirty-one lost that month—more than were being produced by the shipyards.

The personal and professional blows made Dönitz realize that he could not win simply by increasing the number of U-boats. As he later wrote, Germany would have to make radical changes to counter the advantages that the Allies had, and at that moment Germany could not do so. Accordingly, on May 24, 1943, the commander in chief of the Germany Navy ordered his submarines to stop attacking convoys in the North Atlantic and return to bases in France, until new weaponry had been developed that could counter Allied radar, aircraft patrols, and surface escorts.

On June 3, 1943, Dönitz assembled his recently returned submarine commanders. For every three U-boats that had left their berths in France or Germany thus far that year, only one had returned. Moreover, because of the high rate of loss, the age of the submarine commanders had dropped, so that the average age of the U-boat skippers listening to his speech was twenty-three. The commander in chief of the German Navy told them how they had been defeated:

> High aircraft superiority with a ratio of 7:1 and the use of aircraft carriers in convoys closed the air gap in the North Atlantic. But this would not be important if the enemy did not possess a secret detection method. As we do not know what it is, there is no counter offensive. The British discover U-boat patrol areas without our knowledge and go around them. If convoys are still attacked they prevent U-boats from moving forward. At night and in poor visibility there are unexpected attacks on U-boats.

In the next four months, from June to October 1943, not a single Allied merchant or escort ship in convoy was lost to a German submarine. The Battle of the Atlantic was won by the Allies, and in large measure by the Allies' superior scientific and technological prowess.

| CHAPTER NINE |

The Great Shift

While Great Britain, the United States, and Canada were cooperating in using new science-based weapons to subdue the menace of the U-boats, in the winter and spring of 1943 they were conversely refusing to share other military-related scientific breakthroughs, among them those having to do with nuclear research.

After the Fermi pile experiment in Chicago, Conant and Bush told President Roosevelt, "We still do not know where we stand in the race with the enemy toward a usable result, but it is quite possible that Germany is ahead of us and may well be able to produce superbombs sooner than we can." Accordingly, Bush asked for a breathtaking commitment of up to $400 million for the construction of uranium separation and enrichment plants and the production of fissionable material. The president agreed to the plan. A few weeks later Great Britain proposed a joint American-British-Canadian effort on nuclear research, but the United States was no longer eager to collaborate. Conant took satisfaction in writing out tough conditions: that there would be complete exchange of information on design and other factors "only if the recipient of the information is in a position to take advantage of it in this war," meaning that Canada and Great Britain would feed information to the United States but not receive it. The British thought these conditions onerous and stalled the negotiations.

An added reason for the American toughness: the Joint Chiefs of Staff had learned that Great Britain had been forwarding other military secrets to the Soviets for years; the chiefs were aghast, as some of those secrets had consisted of American inventions that were not supposed to be passed on. American fears of Soviet betrayal were realized when an American mission went to Moscow seeking information on Russia's more advanced synthetic rubber manufacturing process but was given a runaround and never obtained the promised information.

The real issue was who would control what patents in the postwar world. In that sense the sharing-secrets problems were an indication of the great shift in the fortunes of war that was taking place, in which the earlier successes of the Axis powers were yielding to an emerging Allied dominance. The change that had begun with Midway, and had continued with the resistance at Stalingrad and victories in North Africa, in the spring of 1943 became more sharply defined by the defeat of the U-boats in the North Atlantic, by the relentless Allied bombing of Nazi-controlled land areas, and in the South Pacific by the Allied retaking of Japanese-controlled islands, one by bloody one.

The host of science-based systems now contributing to this shift were the fruits of years of preparation and organization building and of the collaboration of Allied political and military chiefs with civilian scientists and manufacturers. Emblematic of the extent of coordination achieved in the United States, a pilot plant for new explosives was able to start production just thirty days after a civilian scientist conferred with the manufacturer's chemical engineers; and within a week of the Navy's learning that the Japanese were using radar of a lower frequency than could be picked up by the fleet's receivers, General Electric was able to design, manufacture, and deliver to the Navy fifty new tubes that could detect the lower frequencies. The theme of science workers' conferences held in Great Britain, the United States, and the USSR (linked by telephone and telegraph) was how to achieve even greater coordination. A British professor proposed that "bishops of science," chosen by workers in each field, should have permanent seats in the House of Lords and that science be recognized by Great Britain as one that House's Estates of the Realm.

Such a proposal could not have been made in Germany or in Japan, where the absence of enthusiastic cooperation among scientists, technologists, and their military and political leaders had resulted in few new weapons or other aids to battlefield superiority. The refusal by the Axis countries' militaries to acknowledge that victory might depend in any way on their scientists' work contributed significantly to the ultimate defeat of the Axis.

Germany's rocket research was a case in point. When Albert Speer took over as chief of armaments, one of his first visits had been to Peenemünde, where he witnessed two demonstrations of rocket prowess, one in planes—they performed remarkably well—and the other in missiles. The missile's launch was very impressive, breaking the sound barrier soon after liftoff, but then the missile turned from its designated path and landed less than a mile from the launch site. Such failures should have been expected in a project so complex; had Speer understood that and pushed for full funding anyway, it is likely that the failures would have been encompassed and surpassed—Dornberger, von Braun, and associates had previously proved themselves capable of learning from mistakes. But in June and again in August 1942, after watching a similar failure, Speer felt compelled to report to Hitler that the rockets were not ready for mass manufacture. Scarcely sixty days later, in October 1942, the rocketeers succeeded in sending a rocket up sixty miles to the edge of space and returned it to earth within a few miles of the target area.

Dornberger's address to his troops, marking this occasion, stressed scientific advance and social benefit:

> We have invaded space with our rocket, and for the first time . . . have used space as a bridge between two points on earth; we have proved rocket propulsion practicable for space travel. To land, to sea, and [to] air may now be added infinite empty space as an area of future intercontinental traffic.

But he added, "So long as the war lasts, our most urgent task can only be the rapid perfecting of the rocket as a weapon."

Speer continued to feel that "the rocket still needed considerable development before it could lend itself to mass manufacture." However, he now agreed to use concentration-camp labor on armaments throughout the Reich and to aid the rocket-missile program.

The guided-missile aspect of rocketry was then only one among a half dozen uses of rockets being developed in the Third Reich: the rocket-propelled plane; a land-to-air rocket missile guided toward a target by heat-sensing equipment; several types of jet planes and remote-controlled flying bombs; new torpedoes; a hydrofoil capable of traveling at 50 knots; jet-powered grenades. "We were literally suffering from an excess of projects in development. Had we concentrated on only a few types we would surely have completed some of them sooner," Speer would later write. Typical of the underfunded projects was a torpedo called a "spider" because it remained attached to the submarine by a thin wire as it traveled through the water. The submarine could send instructions through the wire to control the torpedo's path and depth, altering its course to counter the target's evasive actions. Only a handful of technologists were assigned to develop the "spider," and it languished; but wire-guided torpedoes would become the standard for undersea warfare in the 1960s.

Among the rocket-based projects, Speer favored the 25-foot-tall, ground-to-air defensive missile Waterfall, able to carry a 666-pound payload along a directional beam to blast an Allied bomber out of the sky. But Waterfall was a defensive weapon, and Hitler refused to give much thought to anything but offensive weapons.

On April 5, 1943, the Nazis announced to the world that they had found a mass grave containing 14,500 bodies in the Katyn Forest and that they were shocked that Soviet forces could have murdered so many Poles, presumed to be soldiers who had once been in Soviet prison camps. There was no mention of similar atrocities committed by German forces or of the extermination programs in Nazi concentration camps, within a few hundred miles of the Katyn Forest, which had already killed several million people. The German report on the Katyn Forest was true, but it was also sand in the eyes of the West, to counter potential stories about the Nazis' mass killings.

Small items in German-language science publications available to the West spoke volumes about the shift in the war. For example, a curious and telling item about Germany was printed in a Swiss medical review, *Praxis.* It reported that the psychological testing service of the German armed forces had been severely cut back. Previously, the magazine explained, "the service was charged with essential tasks of aptitude testing for the special branches of the armed forces and for purposes of promotion." Prewar testing had gone well beyond the administering of standard questionnaires. Interviews were conducted by teams of psychologists and the ratings were cross-checked to produce a final recommendation on an individual being evaluated. That system had handled 100,000 men a year, but came under pressure to process far more, and its first concession to war was to reduce the time the psychologists spent with each man being tested. The second cutback was in response to a report of a shortage of military psychologists, especially in the Luftwaffe: a new university-degree program for psychologists was instituted, below the Ph.D. level that had previously been required, so that more psychologists could be trained and used in the service. A further curtailment of the testing service two years later, the Swiss magazine suggested, meant either that there were still not enough psychologists in Germany to do the tasks or that draftees who in prior years would not have passed psychological muster now had to be quickly judged acceptable and sent to the front lines.

Articles in the German chemical-industry magazine *Chemiker Zeitung* predicted that after the war (and the presumed German victory) Europe would no longer require foodstuffs and raw materials from the British Empire, since those had been replaced by synthetics and ersatz foods and materials. But reports from the occupied countries showed that this was not so. For instance, Germany still needed Danish farmers to supply significant quantities of meat and dairy products. In exchange, Hitler had accepted an arrangement that permitted Danish Jews to remain in their country, untouched.

Allied magazines such as *Nature* reported that due to the Allied blockade, which was quite effective, Germany was now acting on the principle that the home country would be the last to starve—which, must already be happening, since deficiency diseases were being found among the captive populations in Europe. Rickets, beriberi, pellagra, spontaneous bone fractures, tuberculosis, and nephritis were common. There was a scarlet fever epidemic in Denmark, typhus in Poland, a tuberculosis outbreak in Belgium. Inadequate nourishment had so affected pregnant women and the babies they produced that many mothers now believed a second pregnancy was tantamount to maternal suicide. Rations in the occupied countries fell to between one-half and two-thirds of the minimum recommended for good health; Jews received even less food. German agricultural scientists had insisted on the sowing of calorie-producing crops—cereals and potatoes—in the occupied countries in order to keep the daily caloric intake of the populations above actual starvation level, but there was concern that not enough of the harvest was being kept for seed purposes.

At the end of February 1943 the German-controlled heavy-water plant at Vermork, Norway, was badly damaged in an Allied commando raid; the existing stocks of heavy water at Vermork were destroyed, as was the plant's capacity for making more. The raid was led by the group of Norwegians who had failed in the earlier attempt and had hidden in the forest during the winter, together with more exiled Norwegians from Great Britain who had been parachuted in. The Allies hoped that the Vermork raid would set back the German nuclear effort by a year or two, but in six weeks the Germans rebuilt the apparatus and restarted it by using part of the stock of heavy water that Heisenberg kept in a tub in a basement at Berlin-Dahlem. German military strategists convinced themselves that the Allied raid on Vermork meant the Allies feared that the Germans were ahead in the nuclear race—so that there was no need for Germany to speed up its atomic research.

At an April 1943 conference, Carl Ramsauer reiterated and emphasized the points he had made to the authorities two years earlier: that Anglo-Saxons were doing all of the cutting-edge physics—there were 37 cyclotrons in the United States and only one in Germany—and that the 3,000 German physicists on the front lines would be meaningless in battle, but those 3,000 physicists working in the laboratories might help win the war. Speer was convinced by Ramsauer's reasoning, as was Goebbels when he read a transcript. Their conversion, and the defeat of the U-boats in the North Atlantic by Allied wizardry, finally convinced Hitler in May 1943 that advanced science might provide ways of preventing Germany from losing the war. At Speer's suggestion Hitler established a Planning

Office for Research and placed at its head Walter Osenberg, a former engineer with a torpedo-research lab, whose first action was to extend a special classification, exempting fifteen thousand scientists from frontline duty so they could remain in or return to various laboratories. Radar specialists were recalled from the front lines for the laboratories; six new radar systems emerged from their work in the next six months, but came so late in the conflict that Osenberg would eventually contend that "Germany lost the war because of incomplete mobilization and utilization of scientific brains."

Part of the problem was the quality of the scientific administrators and the information they were conveying to the political and military leadership. Three of the main zoology institutes, for instance, were headed by men widely regarded by biologists as incompetent—Feuerborn in Münster, Lehman in Tübingen, Kuhn in Cologne—and nothing of substance emerged under their leadership. Mentzel, the head of the Reich Research Council, was an even higher-placed stumbling block; for example, on July 8, 1943, he wrote to Göring that although German atomic research "will not lead in a short time towards the production of practically useful engines or explosives, it gives on the other hand the certainty that in this field the enemy powers cannot have any surprise in store for us."[1] Hitler, Göring, and Speer did not know enough science to realize that Mentzel's conclusions were political rather than scientific, or enough to label them suspect because they came from a second-rate chemist who understood little of the complex problems or possibilities of chain reactions.

The one scientific area where the Germans were clearly ahead of the Allies was biological and chemical weapons. In July 1943 a working group named Blitzableiter began looking into biological weapons,

[1]In Japan, at around this time, a high-level committee that included Yokio Nishina and other scientists then working with the military calculated that it would take Japan ten years to make an atomic bomb and that to refine the uranium would require a tenth of the country's electrical energy and half its copper. In addition, their analysis led the conferees to believe that similar requirements would also preclude Germany and the United States from being able to make an atomic bomb.

under the direction of Kurt Blome; the impetus had supposedly come from a report that the Americans were going to drop Colorado beetles on Germany to destroy crops. Here the stumbling block was Hitler, who, according to the minutes of a meeting, "has prohibited all work dealing with the active use of biological weapons." Blome and some of the scientists involved seemed determined to find a reason to do the work anyway. The minutes continued: "The utmost effort must therefore be put into working on defensive measures. It is therefore important to know how the enemy could use biological means; for that reason delivery methods have to be tested." When an expert, Heinrich Klewe, concluded that the evidence showed that the Allies were not considering any large-scale biological weapons, Erich Schumann, head of the Science Section of the Wehrmacht—another old Nazi and an incompetent scientist and administrator—rejected this conclusion. Certain that the Allies were going to attack with biological weapons, he insisted,

> Surely the Führer is not sufficiently informed; he must be briefed once again. We must not watch heedlessly but must also prepare for the large-scale use of biological weapons. In particular, America must be attacked simultaneously by various human and animal epidemic pathogens as well as plant pests. The Führer should be won over for this plan.

So research went ahead, encompassing (in Klewe's words) "plague, typhoid, and paratyphoid bacilli, cholera vibrios, and anthrax spores." The anthrax, released on an island in a lake, succeeded in infecting most of the animals on the island, proving that anthrax would be useful against England. Another well-regarded pest for England was the potato beetle, tested in the summer of 1943 on a field in Germany. These tests and research were carried out under Blome in a facility disguised as a cancer-research institute, on the grounds of a former monastery. Later on, Blome proposed doing human experiments to test the efficacy of the agents. The Wehrmacht refused to have anything to do with this but agreed that it should be done by other branches of the government. Foot-dragging in completing arrangements for such testing served to make further remote the possibility that biological weapons would be ready for use by Germany in the war.

The military fortunes of Germany were shifting so rapidly that in late 1943, when twenty-two KWI directors met with Albert Vögler, most of the meeting concerned plans for evacuating the major research laboratories from Berlin so that they would not be bombed out of existence, as had nearly happened to the KWI for iron research at Düsseldorf in June. One-third of the "uranium club" scientists were relocated out of Berlin or otherwise left the project—Max von Laue took the occasion to retire to Württemberg—but the most important ones and their experiments remained at Berlin-Dahlem, transferring to a newly built underground bunker, and in these quarters they began a full-scale push toward an atomic reaction. Walter Gerlach, a well-regarded scientist who had previously worked on torpedo fuzes, was tapped by Speer to become deputy to the Reich Marshal in charge of atomic physics, replacing Mentzel.

There was mounting pressure for scientists to contribute to the war effort, but the very desperateness of the military situation made it more possible for German scientists to resist that pressure. When organic chemist Karl Ziegler was offered the directorship of the KWI's Institute for Coal Research, he agreed to accept only if he did not have to limit his research to the industrial use of coal but could deal with the much more basic general chemistry of carbon compounds. His conditions were accepted. That basic research would lead to Ziegler's sharing the Nobel Prize for chemistry in 1963.

It was also at this time, late 1943, that the Germans undertook the most massive mathematical enumeration project ever envisioned in the history of the world, one that would go well beyond the Osenberg lists of scientists and facilities to number and register every person, every building, every nut and bolt and piece of machinery and property in the Third Reich.

While the effort to bring German science more fully into the war sputtered and was only partially successful, in the United States more scientists—and more varieties of scientists—brought their talents and training to militarily related tasks. In the spring of 1943, when Grand Admiral Dönitz was scrapping the excess German surface navy and when the German armed forces were stinting on the services of psychologists, American psychologists were becoming more actively involved in military work than ever, helping to staff the newly commissioned battleship *New Jersey*. On board that ship was some of the most complicated and sensitive electronic equipment ever produced—and a crew of 2,600, half of whom were raw recruits and two-thirds of whom had never been to sea.[2] The executive officer of the ship had been working with the Navy's personnel bureau and had asked the NDRC's Applied Psychology panel for help in evaluating and training the crew. Batteries of tests were devised to discern the aptitudes of the seamen and assign them to tasks that made use of their abilities; then the psychologists helped technical experts train teams in the proper use of the radars, fire directors, communications devices and other advanced machinery. The results—better shakedown cruise, fewer transfers, greater efficiency of the crew—were soon the envy of similar ships whose captains had initially been skeptical of the "headshrinkers." In this collaboration the field of psychology also benefited, adding to its store of knowledge about human behavior and how to ascertain information that might predict an individual's future behavior. For psychologists in the United States,

[2]Most of the Navy's recruits were literate; examinations for the Army revealed that as many as 20 percent of draftees could not read well enough to understand written orders. Schools were established at base camps to teach reading, writing, and elementary math to hundreds of thousands of soldiers.

Canada, and Great Britain, the war meant a chance to evolve their science from "soft" to "hard," able to measure and evaluate results as was done routinely in physics, chemistry, and biology.

Despite the presence in Santa Fe of some of the best scientific minds in the world, advances in the science of nuclear physics were not even expected when, in March 1943, the Manhattan Project took up temporary residence there while new buildings to supplement the boys' school at Los Alamos were under construction. The recruits from Chicago, MIT's Rad Lab, Berkeley, and abroad were under no illusions that they had come to find new knowledge; their task was practical, fashioning a bomb under the leadership of J. Robert Oppenheimer of the University of California, who had briefly trained at Göttingen in 1926. Oppenheimer had been chosen as chief despite his history of left-wing politics—he had belonged, he wrote on an application, to "just about every Communist front organization on the West Coast." The Army had at first rejected him for any sensitive post; Bush, Conant, and Lawrence pushed for his appointment because he was a first-rate scientist and determined to build a bomb. Oppenheimer's resolve would be matched in the bomb project by the enthusiasm and expertise of émigré physicists Rudolf Peierls, Hans Bethe, Edward Teller, and Viktor Weisskopf, who were, as Bethe later put it, "desperate to do something—to make some contribution to the war effort" to defeat the Nazis who had taken over the country of their birth and training. Émigré Felix Bloch, recruited by Oppenheimer, later recalled that he went to Los Alamos even though "I wasn't sure, first of all, what I would do there and whether I could really live in this military atmosphere. But nevertheless I felt it was my duty at least to try." The security measures proved too much for Bloch; he lasted only a few months at Los Alamos, then went to Harvard to work on radar countermeasures.

In initial lectures for the small Los Alamos staff, the object of the project was crystallized: "to produce a practical military weapon in the form of a bomb in which the energy is released by a fast neutron chain reaction in one or more of the materials known to show nuclear fission." For some scientists the details of these lectures were revelatory, because compartmentalization of information had previously prevented each set of scientists from knowing what other sets were doing. When an engineer suggested the bomb should implode, even such a first-class scientist as Richard P. Feynman ridiculed the notion; soon, however, consultants John von Neumann and Edward Teller worked out the theoretical and mathematical parameters of implosion, and attempts to construct an imploding bomb proceeded. James Tuck, the Oxford-educated friend of Reg Jones who had for years served as Cherwell's adviser, contributed the idea of an explosive "lens"—succesive focused shock waves—to help trigger the plutonium core.

In May 1943 many scientists and administrators of the Los Alamos endeavor still had doubts about success. Moreover, James Conant wrote just then in a secret history, "everyone concerned with the project would feel greatly relieved and thoroughly delighted if something would develop to prove the impossibility of an atomic explosive"; they would much rather the experiments prove the feasibility of an atomic reactor that could provide the world with "atomic energy for power."

In April 1943 American cryptographers learned from intercepted Japanese coded cables that Admiral Yamamoto, architect of the attack on Pearl Harbor and the preeminent Japanese naval tactician, would soon travel to a small island by air. President Roosevelt, Admiral Nimitz, and other high officials agreed that going after Yamamoto was worth the risk of exposing that the Japanese military codes had been broken. The bomber in which Yamamoto rode was shot down by American fighters, killing him, and even after the event the Japanese Navy continued to use their codes, which they believed were unbreakable. Historians now mark the assault on Yamamoto as the moment when Allied forces turned from defense to active counterattack in the Pacific.

Around this time the British finished deconstructing the codemaking machine found in North Africa, and their mechanism was used to build a small, semielectronic machine on the same principle. It still took days to decipher German Fish messages, however, so T. H. Flowers of the British Post Office Research Station—which had long been interested in the perforated-tape mechanism at the heart of the teleprinter—undertook to reproduce and enlarge the design on a grand scale, 1,500 valves rather than the 100 in the smaller version, a new machine to be called the "Colossus." It was the world's first true computer.

The Allies planned to land on Sicily and then work their way north through Italy. Partly to distract the German military from this theater of operations, but mostly to dishearten the German populace and its military leaders, the Allies launched increasingly accurate and ever-larger bombing raids on German war-making sites and population centers. In Allied aerial attacks—the planes and bombs guided by sophisticated radar, the explosive power magnified through research into the chemistry and physics of the charges, the bombing patterns selected for greatest effect through statistical analysis—science and technology were applied solely to the task of achieving the greatest possible terror and destruction.

The escalating bombing over Europe included the climax of a war of radars and attempts to jam their signals. To obtain data on the German radar known as Liechtenstein, mounted in night-fighter planes, a British search aircraft flew slowly in front of one such fighter, allowing itself (and its crew) to be hit repeatedly in order to gather information; the most wounded crew member was pushed out of the plane in a parachute over Great Britain with the logbook of observations tied to him, in case the rest of the crew and the plane were not able to limp home. The information was crucial to the development of the jamming technology known as "Grocer." Another developing jamming technology was "Moonshine," a pulse repeater designed to simulate the approach of a nonexistent bomber squadron, causing the scrambling of German fighters in the wrong direction. A third was a line of jamming transmitters called "Mandrel." The United States contributed "Ferret," a radar-searching plane, the first to be sent on "electronic reconnaissance" missions, and the powerful jamming transmitter known as the "Tuba," which after the war would be used as the basis for broadcasting television signals. Various Allied jamming measures known by such evocative names as "Shiver," "Cigar," and "Airborne Cigar" were arrayed against radio transmissions from ground to pilot or between airplanes. When the Germans switched to voice instructions over the radio to their pilots, German-speaking Britons intervened with bogus instructions; when the Germans countered this with female voices radioing instructions, Reg Jones replied with more authoritative-sounding male voices, to ensure that "confusion was immediately restored."

For several years Cherwell had pushed particularly for the development of the aluminum chaff known as "Window," to be dropped by an aircraft above a bombing target to bewilder enemy antiaircraft-battery radar. A pound of scattered aluminum strips would reflect ground-based radar in such a way as to give the illusion that it was a plane. Astronomer Fred Whipple and associates figured out that the most effective radar echo would be produced by very thin strips, one-half the size of the wavelength of the radar to be jammed.[3] An added bonus for the jammers: on the reverse (paper) side of the strips, propaganda messages could be printed.

Chaff was such an obvious development that German, British, and American laboratories all had studied its possibilities. During

[3]The reason the strips could be so precisely calculated—and the total weight of the aluminum required could be so low—was that the Germans had bunched most of their radars in the 550-to-570 megacycle range, a very narrow bandwidth.

1942 and early 1943 both belligerent sides held back on using chaff, each believing that the other might gain an advantage if its use could be predicted. In the spring of 1943 Cherwell's advocacy of chaff and his ability to influence Churchill forced a decision on its use. After a debate in the War Cabinet, during which those opposed to chaff made the absurd contention that the RAF would need a hundred tons of aluminum to upset German radar but that the Germans would need only one ton to disrupt British radar, Churchill issued the order: "Open the Window."

Following a disastrous Allied raid on Leuna, in which 125 aircraft were lost, Window was used in conjunction with the most massive Allied air raid of the war so far, the July 1943 bombing of Hamburg. Window's contribution to the Hamburg raid was to reduce the accuracy of the enemy's antiaircraft guns. British bomber losses were cut from the usual 5 percent to one-half of 1 percent—from 125 in the raid on Leuna to 12 over Hamburg. So many Allied incendiaries were dropped over Hamburg that observers in planes saw not many fires but one huge, all-enveloping fire that destroyed eight square miles of the city—an area larger than the island of Manhattan—and sucked the air out of underground shelters, suffocating the inhabitants; 45,000 people died. It was no wonder that the British dubbed the bombing of Hamburg the first phase of Operation Gomorrah. It was terror bombing of the sort that Allied civilians, scientists, and even military professionals had decried when Germany had done it in the past. Whatever scruples British and American military planners had once professed about terror bombing seemed now to have been suppressed. The massive Allied bombing of German cities soon exceeded the scope and ferocity of what the Germans had done to Guernica, Warsaw, Coventry, and London.

To the British public the bombing of Germany was portrayed as targeting military and industrial objectives, with civilians killed solely by accident. Only by portraying the bombing in this way, a high-level air ministry memo said, could the authorities "satisfy the inquiries of the Archbishop of Canterbury, the Moderator of the Church of Scotland, and other significant religious leaders whose moral condemnation of the bombing offensive might disturb the morale of Bomber Command crews."

Goebbels, Göring, and other German leaders wrote in their diaries and memos sober assessments of the terrible loss of life and property in Hamburg and worried for the future of the Reich. So did General Martini, the radar expert, and he added that "the technical success of this action must be designated as complete. By this means [chaff], the enemy has delivered the long-awaited blow against decimeter radar sets both on land and in the air." In response to the raids on Hamburg and other cities, Hitler ordered many fighter planes and 8.8-centimeter antiaircraft guns to return from Russia to Germany; the "88s" were the most effective antitank weapon the Nazis possessed, and they had proved important in the huge tank battle near Kursk, but Hitler now brought them back to protect the homeland. Soon the returned complement of FW 190 fighter planes and 88s served to somewhat counter the Allied radar advantage. The Germans also sped up the production of a Doppler-effect-based system that could distinguish between chaff and airplanes; on August 17, when 376 American Flying Fortress bombers tried to obliterate the ball-bearing factories at Schweinefurt, the new antichaff system was waiting, along with the FW 190s, armed with new R4/M rockets; the combination shot down 60 of the bombers and prevented the destruction of most of the factories.

The 10 percent increase in the ability of the Allied forces to sink U-boats tipped the balance of the Battle of the Atlantic in favor of the Allies. Similarly, Window and other jamming techniques cut Allied planes losses over Germany and added another percent or two to the Allies' growing edge in the war. The percent of the overall increase in the ability of the Allies to wage war attributable to science and technology cannot be accurately quantified, but the contributions were numerous and cumulatively significant. Not all were as terrifying as the atomic bomb or as marvelous as radar; many were mundane. For instance, Canada had laboratories that tested various military-related equipment: shock and vibration machines subjected electrical junctions and sensitive instruments to extensive stress, and the results led to fewer machinery failures; ammunition was redesigned so that it would not blow up in gun barrels; the armoring properties of materials were evaluated so that the most appropriate could be installed in airplanes; optical equipment was refined and calibrated; the composition of fuze powder was adjusted. A rubber laboratory improved surgeons' gloves, gas-mask components, ground-cover sheets, artillery and tank parts, crash and steel helmets. A leather lab and a textile lab worked to prevent boots from cracking, to reduce weathering of canvas duck, to flameproof cotton uniforms, to substitute plastic for metal. A biology committee looked into ways to restore dehydrated foods to edibility, to better preserve bacon on transoceanic voyages, to treat eggshells to prevent deterioration at ordinary temperatures—unable to achieve this last goal, they recommended that all eggs be shipped as powder. Each project might have increased the efficiency of the Allied fighting forces perhaps one one-thousandth of a percent—but there were hundreds of them. Australian specialists concentrated on matters of food and medicine supply, raising the efficiency of growing crops and livestock and facilitating the extraction from native and imported plants of atropine, digitalis, quinine, and strychnine. They investigated shark oil and seaweed, rich in vitamins needed by soldiers in jungle environments; these studies were considered no less important to the war effort than those investigating ethylene as a starting material for the synthesis of organic chemicals.

The U.S. R&D effort was the most massive. There were dozens of laboratories within American military installations, supplemented by eighty contracts awarded each year to universities and double that number to private companies—by 1943, $100 million a year was being spent specifically on military-related research and development. Chain and rope were tested in the Boston Navy Yard, rubber and paint at Mare Island, California, and torpedoes at Newport; 5,000 chemical analyses a week were done in Pittsburgh. Under the direction of Warren Weaver and of Hunter College mathematics professor Mina Rees, mathematicians at New York University (working with Courant), Brown, Harvard, Columbia, and other universities tackled mathematical and statistical analyses of shock waves, jet-engine stress, underwater ballistics, and similar problems, obtaining data directly relevant to the design of offensive and defensive weapons.

Katharine Burr Blodgett, an American who was the first woman to receive a doctorate in physics from Cambridge University and the first female scientist hired by General Electric, put aside her studies on thin films to work on the de-icing of plane wings and the construction of a chemical smoke screen used to shield troops from enemy eyes. Walter Ladenburg, an émigré physicist, one of the first men to have proved Einstein's theses, put aside his research at Princeton to join the Army's ballistics lab, for which he developed a flash suppressor for rifles. Percy Bridgman, who would win a Nobel in 1946, worked on the plastic flow of steel under pressure, to strengthen armor plate, before he went to the Manhattan Project. One future American-citizen Nobelist who was not utilized well was Maria Goeppert-Mayer, the Göttingen-trained émigré, who during the war was relegated to minor work in Harold Urey's laboratory.

American scientists and military men initially brushed aside British requests to work on large flamethrowing engines, perhaps believing that burning people to death was more repugnant than killing them with bullets. But after American armed forces experienced great difficulty in dislodging Japanese defenders from underground bunkers on Kwajalein and Tarawa, with consequent loss of American life, scientists participated in concocting gels for and perfecting the mechanisms of portable flamethrowers. These were tried out at ranges in Virginia and Florida where elaborate models of the

Japanese fortifications were replicated. This research also produced the "jellied gasoline" used as an incendiary in air raids that destroyed most of Tokyo. Merle Tuve would later recall that long after the war, Vannevar Bush would wake up screaming from nightmares caused not by visions of the havoc wrought by the atomic bomb or the proximity fuze, but by the "jellied gasoline" produced under his direction that had burned to death so many Japanese civilians.

When basic research on incendiaries began, there seemed to be a need for deliberately obscuring its purpose. A project at Columbia was blandly described as "a study of the corrosion of iron by copper chloride solution," so that no outsider would understand that they were perfecting the chemical mechanism of an incendiary device called a "Pencil," widely used for sabotage.

One of the largest-scale and most secret projects involved the development of new poison gases. Compound 1120, four times as lethal as phosgene, was invented by American and Canadian researchers, and a pilot plant was set up to manufacture it. The compound was lethal, colorless, and heavy. It was also believed to be odorless, but since no one could have breathed the amount of gas necessary to decide that question and stayed alive, it remained an unknown point. To prevent the formula from being stolen, it was never written down in any document that could be taken away from secure premises.

Two hundred committees of scientists oversaw scientific investigations in such fields as aviation medicine, gas chemistry, and jet propulsion; proper nutrition and health care for workers in factories; and even into such far-out ideas as whether soldiers would accept reindeer meat and low-nicotine burley cigarettes. The British estimated that the United States spent five to ten times as much per citizen on scientific research as did Great Britain and that one result would be American scientific and technical superiority in the postwar world.

In the United States Congress, committees headed by Harry Truman, Harley Kilgore, and Claude Pepper in the Senate and by Wright Patman in the House of Representatives held hearings on the inadequate use of science and technology by the government—for instance, in the production of synthetic rubber and in the utilization of specialists on the 600,000-person National Roster. A Kilgore-Patman bill called for science mobilization on an even grander scale than had already been accomplished. Debate over putting all of science and technology under a single supervisory body continued throughout 1943. One side called it a "Magna Carta for science," the other, regimentation that would enslave science and industry as they were currently enslaved in Axis countries. The bill was eventually withdrawn because of scientists' opposition, but the discussion about the need to direct science toward national goals lingered on.

One aid to battle that the Axis belligerents lacked and that the Allies were beginning to have, in mid-1943, was the "wonder drug" named penicillin. A year earlier there had not been enough of it in the world to treat more than a few patients. Although Alexander Fleming had discovered penicillin in 1929, a decade elapsed before H. W. Florey, N. G. Heatley, and Ernest Chain at Oxford had concentrated and dried the mold product into a substance that could be injected into a human being without adverse side effects. At the start of World War II, penicillin was little more than a laboratory curiosity and could not be manufactured in quantity. When a Rockefeller grant brought Heatley and Florey to the United States in July 1941, they took their problems to a group of researchers in Peoria, Illinois, at the Northern Regional Research Laboratory of the Department of Agriculture. At that moment no one knew what strains of mold best produced the type of penicillin that could tame bacterial infections in human beings, or what was the most favorable medium for growing the mold, or the best method of extracting penicillin while preserving its potency and purifying it, or how to manufacture penicillin in the vast quantities required. A team of fermentation specialists went to work, and four months later they had enough data to convene a meeting of pharmaceutical companies and tell them details of what had become a closely guarded national secret. Yields from the mold had been increased between ten- and twentyfold, and a good medium for growing it had been found—"corn steep liquor," a by-product of the wet-corn-milling industry, mixed with brown sugar. But in June 1942 there was still only enough penicillin to experimentally treat ten patients and enough for 100 patients by February 1943.

Then a large-scale testing opportunity presented itself: 500 GIs, wounded and disease-infected on the Pacific Islands, brought for recuperation to a hospital in Brigham, Utah. They were given penicillin in controlled conditions; as a later official report put it, the results "were so striking that they could not fail to convince the medical corps of the armed forces and others that the drug possessed military importance of the highest order." Penicillin cured disease conditions that had proved resistant to all other known drugs, particularly the staph and strep infections that had been the scourge of hospitals, as well as pneumonia, meningitis, gangrene, and the venereal diseases gonorrhea and syphilis. Production of penicillin in mass quantities was taken over by the War Production Board and private companies, and enough was produced in the United States and in Great Britain to have a major impact on the rates of recuperation in Allied hospitals from late 1943 through the end of the war.

Since the Middle Ages the most dangerous disease faced by armies had been typhus. During and just after World War I, lice-borne typhus killed 135,000 people in Serbia, 250,000 in Rumania, and 3 million in Poland and Russia. No known drug—not even penicillin—could kill the microbes that caused the disease; typhus could be prevented only by thorough bathing, frequent haircuts, and the steaming of clothing. Crowded conditions, hunger, and cold—which made people not want to change their clothing frequently—bred typhus in epidemic proportions. Russian scientists in the 1930s tried a grotesque way of developing a typhus vaccine. In order to grow enough lice to grind up into vaccine, Soviet technicians scalded the corpses of people who had recently died, removing the skin and then stretching it over shallow dishes of blood so that lice could feed and breed. European scientists tried infecting mice and preparing vaccine from their lungs, but that vaccine caused allergic reactions in many human beings. Finally Herald Cox, an American scientist with the Department of Agriculture, was able to grow microbes on chick embryos in sufficient quantity to make what was believed to be a safe vaccine.

Cox's antityphus vaccine had never been field-tested when, in November 1942, it was given to all American and British troops heading for North Africa. A typhus epidemic was then raging in the area, with 77,000 reported and an estimated half million unreported cases. In many months living in North Africa, only 11 Allied soldiers came down with typhus, and none died from it. Then the problem became how to protect prisoners of war and neutral civilians; there was not enough vaccine, so preventive methods had to be used. But Muslim women refused to take off their clothes and be disinfected, which was the only sure way to rid them of lice. An ingenious solution was found: the insecticide DDT, already becoming effective in the South Pacific against malaria-bearing mosquitoes, was made into a louse powder and sprayed under their clothes with a flit-gun. DDT spray was soon being sold on the black market in Algiers, a sure sign of its success.

Another important medical innovation came about in the course of trying to treat shock, a major cause of death after a soldier had been wounded and had lost blood. In the 1930s it had been thought that blood plasma—the fluid left behind when red and white cells are removed from whole blood—could adequately substitute for blood, but battlefield conditions showed that it could do so only in the first stage after the wounding, and that for full recovery, human bodies needed red blood cells to carry oxygen throughout the body. American and British civilian authorities, military staffs, and medical corps set up systems for delivering plasma to the wounded in the field and at frontline hospitals and for bringing whole blood to the larger hospitals. Meanwhile, laboratories of Harvard, MIT, Columbia, Wisconsin, and other universities undertook research on blood substitutes and on the ways in which blood components acted in the body. MIT was able to radioactively tag certain blood cells and track them through the body—and also to evaluate methods of storing, transporting, and transfusing red cells. The studies determined the best preservative, an acid-citrate-dextrose (ACD) solution.

Enough whole blood for the wounded, especially for those in shock, could not be obtained from their fellow soldiers, so blood had to be shipped from the home countries for use in the Italian campaign. Technologies were rapidly developed for the mass extraction, separation into components, storage, and transport of more than 10 million units of blood, most of it in ACD solution. Later on in the war every invasion of enemy territory undertaken by the Allies was done by armed forces that carried whole blood with them into the battlefields.

Innovations in transfusions used in conjunction with hospitals set up near the front—themselves important contributors to preventing death from battlefield wounds—and with techniques learned from experience—such as not to make sutures in front-line hospitals, only in more sanitary ones[4]—saved tens of thousands of lives. The Allied hospital death rate for the wounded was 3 percent—that is, 97 percent of the injured or diseased who reached those hospitals survived. The better survival rate of Allied military personnel contributed to the Allies' increasing edge over the Axis, whose battlefield and hospital death rates were much higher.

---

[4]The sutures were a scientific advance, too, made by a new method of extracting collagen from slaughterhouse waste products.

In the spring of 1943, as the physicists and engineers were taking up residence at Los Alamos to produce the weapon that would end the war, as penicillin was undergoing its first big test, as Allied sub-hunting technologies were chasing the U-boats from the Atlantic, and as the next generations of war-related weapons and sensing devices were coming to fruition in the laboratories of the Allies, the notion that the war would sooner or later end in an Allied victory was already in the air. It occasioned a long article by Waldemar Kaempffert of the *New York Times* on how the science of the war would affect postwar civilian life. The emphasis was on material progress, not on destruction. Light metals were one focal point, plastics another; both would contribute to major changes in transportation and communications: "We shall think of distance in terms of time rather than of miles," Kaempffert predicted, and the "extraordinary developments in short-wave radio brought about by the necessities of war" would banish isolation, eventually result in connecting us all at every moment, and produce ubiquitous television that would transform the movies, radio, and political campaigning. "With each advance in chemistry, engineering and invention," Kaempffert concluded, "we must expect more social adaptation to the machine. It is the price that we must pay when the luxuries of yesterday become the necessities of today."

| CHAPTER TEN |

Vengeance Weapons

On May 24, 1943, the day Dönitz ordered the submarine retreat from the North Atlantic, Dr. Josef Mengele arrived at Auschwitz. A former assistant to Otmar von Verschuer, the leading eugenicist, Mengele had served in the Waffen-SS until 1942, when after being wounded he took a position in Berlin that oversaw the "medical experiments" in the concentration camps. Verschuer then encouraged Mengele to apply to Auschwitz, where he could have the opportunity for research that would advance the field of race biology. At Auschwitz, Mengele immediately began torturing and killing cripples, twins, and young women, all in the service of determining how best to breed a master race. After the deaths of the twins, he would dissect them and send their blood, pairs of eyes, and other body parts—labeled as URGENT WAR MATERIAL—for analysis at Verschuer's KWI at Berlin-Dahlem. As his regular Auschwitz task,

Mengele made decisions that resulted in the immediate deaths of thousands of people arriving at the camp. His associate, another medical doctor, Karl Clauberg, shortly reported to SS leader Himmler that he had worked out a method for sterilizing a thousand women in a single day. Another physician at Auschwitz, Fritz Klein, later explained his actions there by saying, "I am a doctor and I want to preserve life. And out of respect for human life, I would remove a gangrenous appendix from a diseased body. The Jew is the gangrenous appendix in the body of mankind." Each week at Auschwitz thousands of Jews, Gypsies, Russian and Polish prisoners of war, local partisans, and political dissidents were killed and their bodies cremated. Similar mass murders were being carried out at other concentration camps. If Nazi science and technology failed in many other endeavors, it was superbly efficient at executing people and reducing their bodies to dust.

Many Third Reich scientists had still not been put to work on projects of military significance. Some had been assigned meaningless ones. Himmler's strong dislike for flies translated into directing an SS scientist at Dachau to spend more than a year investigating the breeding of a natural parasite to eat flies and mosquitoes. Only later was Eduard May commanded to find ways in which insects could carry deadly diseases into enemy territory.

At Peenemünde, late in May 1943, the Long-Range Bombardment committee met to review two projects, the Army's A-4 rocket and the Fi-103, the Luftwaffe's jet-propelled, pilotless flying bomb. Speer, Milch, Dönitz, and representatives of the other services watched demonstrations and tried to decide which weapon would have the greater probability of success, the "ballistic missile" rocket or the jet-propelled "cruising missile." Both performed reasonably well in tests, though not perfectly. Dornberger contended that it would be a mistake not to forge ahead with both programs, as they were complementary and would both be needed to defeat the Allies.

His recommendation won full funding of both projects and a promotion to major general.

A month earlier a British reconnaissance plane photographed on that Baltic peninsula an unknown piece of ordnance ready for launch. Information from various sources, including a surreptitiously taped conversation between captured German officers, indicated that various long-range weapons might be in the testing stages. But Cherwell had concluded that a long-range rocket was a technical impossibility. That was also the judgment of Great Britain's rocket expert, Alwyn Crow, designer of moderately effective short-range antiaircraft rockets. Extrapolating from his own solid-fuel design, Crow calculated that to reach London a rocket would have to weigh seventy tons and would need a guidance and control system beyond what was currently feasible. He dismissed the idea of a rocket using liquid fuel, even though Isaac Lubbock, a Shell Oil engineer, had been successfully experimenting with liquid rocket fuel.

On the recommendation of the service chiefs, Churchill asked his son-in-law, Duncan Sandys to chair a committee to look into the matter. Sandys was a member of Parliament, a pilot, and had done some work at Britain's own rocket-development station; he had also recently completed a first-rate report to Churchill on deficiencies in the development of jet planes. Sandys's investigative group, Bodyline, commissioned aerial surveys of the Peenemünde peninsula and interpretation of the photographic evidence by scientists to estimate how far the rockets might be able to travel. The scientists' group was headed by Nevill Mott, a Rutherford-trained mathematical physicist at Operations Research.[1] Within two months Sandys had accumulated enough aerial evidence of what was being built at Peenemünde and confirming information from Reg Jones's sources, as well as a report on liquid-fuel possibilities from Lubbock, to recommend that

[1]Mott would share the 1977 Nobel Prize for physics.

the German rocket operations be thoroughly bombed. But Cherwell still disagreed, and a cabinet debate raged.

While the decision was pending in London, at Peenemünde Dornberger launched two rockets in the presence of Himmler; one was a spectacular failure that doubled back on the launchers, dug a hundred-foot hole, and rattled windows 2 miles away; and the other was just as spectacular a success, traveling 147 miles over the Baltic before coming down. Himmler and Speer had been at odds over the rocket program, but both now recommended it to Hitler.

On July 7, 1943, Dornberger and von Braun traveled to the Führer's Wolf's Lair retreat to show color film of the tests and models of the rockets to Hitler. The German leader was preoccupied with current matters: in the Kursk salient in the USSR, the largest tank battle in history was being waged; and it was clear that the Allies were about to invade Sicily. But when the rocketeers showed their film, the mighty power of the launches entranced Hitler. He insisted on having 2,000 rockets produced each month, and the cargo of each increased from 1 to 10 tons of explosives. "What I want is annihilation—annihilating effect," Hitler said, brushing aside Dornberger's objection that there was not enough liquid-oxygen rocket fuel in Germany to accomplish that goal. Hitler even apologized to Dornberger for not realizing earlier how important the rockets could be to achieving Germany's final victory. Hitler told Speer how impressed he was by the thirty-one-year-old von Braun—so young to have accomplished such an amazing technological feat; Hitler likened von Braun to Alexander the Great and Napoleon, who had achieved greatness when they were in their twenties.

Hitler's compliments were accompanied by a pledge of virtually unlimited resources for the rocket missile program—resources on a scale comparable to what the United States had committed to the atomic bomb project, though composed of far different components, such as slave labor and money that might otherwise have gone into food for the population.

A few days after Hitler saw the Peenemünde film, in the Kursk salient the tide of battle turned against the Germans.[2] The Allies invaded Sicily; on July 25 Mussolini was deposed as head of the Italian state. Within the week the Allies bombed Hamburg to near oblivion. These events solidified Hitler's desire to use the new rocket missiles as "vengeance weapons"; soon he was telling the German public on the radio that revenge for Great Britain's dropping of bombs on the Reich's homeland would be forthcoming. In accordance with their new missions, the pilotless flying bomb was renamed the V-1 and the huge rocket missile the V-2.

In July, Hitler also approved the manufacture—and misunderstood the use—of another advanced technological weapon system, the jet plane. After years of holdups and having its development shortchanged by inadequate resources, Messerschmit's fifth version of the jet-propelled Me 262 was finally judged satisfactory and ordered into production. Able to travel at 540 miles per hour and with a range of 650 miles, it could fly rings around any propeller-driven aircraft. Yet Hitler decreed that it be manufactured only as a long-range bomber; his advisers had to plead to get him to agree to a fighter version. Hitler also thrilled to a prototype of the Me 163 Komet, an even more unusual departure from the normal airplane than the jet. The Komet was rocket-powered, lacked a horizontal tail, and had a very short and compact fuselage. Catapulted into the air by a trolley (similar to that which launched the V-1s), it was to land on a "sprung skid" underneath the plane, because it had no landing gear. The Komet was very powerful, able to reach 623 miles per hour, but it had a range of only 62 miles—about 8 minutes in the air; during those minutes, however, the rocket plane could fire its machine-gun cannons with devastating effect on formations of enemy bombers.

On the Allied side a British jet plane was also nearing completion—spurred by that Sandys report on its undue neglect. Called the

[2]Information on the German plans at Kursk, deciphered by the British, was sent to the Soviets by a spy at Bletchley, John Cairncross.

RAF Meteor 1, this jet was designed expressly as a fighter, and it had survived various tests, crashes, and dustups within the Air Force establishment over the suitability of nonpropeller planes. It was slower than the Me 262, able to manage only 480 miles per hour, but production chiefs now told Churchill that he could have some of the Meteors ready in less than a year, for use in the invasion of France.

In early August 1943 the British cabinet finally resolved the debate over what to do about the potential menace at Peenemünde: destroy the sites, even if the precise nature of what was on them had not been ascertained. A massive bombing raid was authorized for the night of August 17–18—500 bombers carrying 4,000 crewmen and enormous amounts of explosives. To draw the Luftwaffe's attention, some bombers aimed for Berlin, 120 miles to the south, while the bulk of them hit the Peenemünde peninsula, where they crushed nearly all the apparatus and development labs of the V-2 rockets, though due to lack of information they did not significantly damage the V-1 facilities.

The raid was estimated to have set back the V-2 program about six months. To deceive the Allies about the progress of rebuilding at Peenemünde, the Germans, at Himmler's suggestion, left many of the destroyed structures unrestored while concealing their replacements. Within ten days of the raid manufacture of the V-2s was transferred to an underground facility near Nordhausen, in the Harz Mountains, mineshafts that had once been considered as a storage place for poison gases. Hitler also authorized work on the "Laffarenz" project, in which a container holding three V-2s in horizontal position would be towed across the Atlantic by a submarine, then righted, set upon a platform, and fueled so they could be fired at American seaboard cities.

In August 1943 the Allies published a report accusing the Nazis of "a deliberate program of wholesale theft, murder, torture and savagery unparalleled in world history," one that had executed 1,702,500 human beings. At the time the figure seemed so high as to appear incredible; later documentation would reveal that the actual death toll at that point in time was much higher.

At the end of August 1943 Churchill, Roosevelt, and advisers met at Quebec to chart Allied strategy for the coming year. The major military decisions taken were to pursue the campaign in Italy northward and to schedule the invasion of northern France for the early summer of 1944. Two other matters, which concerned science, were discussed at the conference, but not with Cherwell, who had taken ill; rather, Churchill's science adviser on this trip was J. D. Bernal, whose openly Communist views seem not to have precluded him from taking part in discussions at Quebec about progress being made toward the construction of an atomic bomb.

The British and Canadians had realized that their countries simply did not have the resources to produce a bomb, while the United States did, and they were now willing to accept Conant's draconian conditions for participation in the research. At the conference Churchill and Roosevelt agreed that Great Britain and the United States would pool their scientific talent to develop the bomb quickly, would never use the bomb against each other, and would not to use it against third parties without prior consultation with each other. After this conference most of Great Britain's nuclear research was transferred to the American continent, along with many of its scientists, some to places like Oak Ridge and Hanford, others to Canadian labs, and some to Los Alamos. Churchill agreed to go forward with an American-led atomic bomb project, he told Canadian Prime Minister Mackenzie King, not only to prevent the Nazis from making an atomic bomb before the Allies had one but also because he "did not want the Russians in particular to get ahead with the process."

However, it is likely that what Churchill feared—a Russian atomic bomb—began at Quebec. Historians now believe that after the conference Bernal deliberately or carelessly conveyed the vast importance of "Tube Alloys" and "Manhattan Project" undertakings to the Soviet Union, through his flatmates Guy Burgess and Anthony Blunt, both then spying for the USSR. It was shortly after the conference that a Russian spy ring was established in Canada, with orders to steal secrets from Canada's atomic-bomb project.

The second "scientific" matter of the Quebec conference was the Habakkuk, Geoffrey Pyke's harebrained scheme to build a floating airfield out of sawdust-reinforced ice; it was this notion that had brought Bernal to Churchill's side for this trip. At the conference Mountbatten wheeled a block of Pykrete into a room at the Château Frontenac and, for the benefit of the Allied Chiefs of Staff, fired two bullets at it. One ricocheted and nicked Admiral King. He recovered, but the Americans vetoed Churchill's proposal for a joint task force, saying only that they would investigate Habakkuk's possibilities on their own—a polite way of burying the idea.

That summer of 1943 Hitler's patience with Denmark thinned as strikes, sabotage, and defiance of Nazi edicts undermined the arrangement by which Denmark had provided large quantities of food in exchange for having some self-government and being able to protect its 7,000 Jews. On August 29, Nazi forces reoccupied Denmark. Shortly thereafter Niels Bohr and his family were advised to escape to Sweden in a small boat. The Bohrs did so, and then the eminent physicist convinced the Swedish king to make public Sweden's offer to care for all of Denmark's Jews; the publicity helped Danish Jews to cross to Sweden. Bohr's persistence and his willingness to assert his influence in Sweden were a key element in the Danish Jews' survival. Bohr then decided to accept a long-standing British offer from Chadwick and Cherwell to come to England to help with certain "problems" in nuclear research. Upon his arrival there, Bohr cleared up one ancillary mystery left over from Lise Meitner's 1938 telegram about the Bohrs and "Maud Ray Kent," from which the name of the MAUD Committee had been taken; Bohr explained that Maud Ray of Kent had once worked for his family, and he had simply been trying to send her a message that they were alive.

In September 1943 the Italian government surrendered to the Allies. The German military command was incensed and took retaliatory measures, among them using an almost-untested device, an SD-1400 gliding bomb. Dropped from a plane, it sank the Italian battleship *Roma* as she steamed toward an Allied-controlled port. The bomb had an armor-piercing warhead that enabled it to go through decks and explode within the ship, doing terrible damage.

German rocket-powered short-range missiles were fired against British ships patrolling the Bay of Biscay, sinking several, which contributed to Admiral Dönitz's willingness to order U-boats back into the North Atlantic. As he had promised his submariners when he had taken the U-boats out of operation the previous spring, the returning submarines were improved machines of war—fitted with snorkels, as well as with better torpedoes, sonar, and radar. Speer and Dönitz had also taken steps to ensure that the next generation of replacements for these U-boats would be even better: Type XXI submarines, with increased battery capacity and greater speed, which were to be used until the more advanced ones—fueled by the oxygen produced chemically from hydrogen peroxide and able to make 21 knots an hour underwater—were expected to be ready, in mid-1944. Speer's office put together an ingenious prefabrication manufacturing design that enabled a Type XXI to be built in two months, whereas the older subs had taken eleven months. Just then Speer and Dönitz often asked themselves "why we had not begun building the new type of U-boats earlier. For no technical innovations were employed; the engineering principles had been known for years. The new boats, so the experts assured us, would have revolutionized submarine warfare." The answer to their question, though they could not say so openly at the time, was that Hitler, Göring, and Raeder had believed that the war could be won without improving the submarines—which the hierarchy always considered as defensive rather than offensive weapons.

American, British, and Canadian labs had devised countermeasures to foil the new German torpedoes: "Foxer," a 3,000-pound hollow tube with holes in it that made it sound like a pitch-pipe when dragged through the water—no ship towing a Foxer was ever sunk by an acoustic homing torpedo; sound-beacon decoys; explosive "popcorn" and "pepper" devices that overwhelmed enemy torpedoes' sensitive listening apparatus; random-frequency and broadband radiowave generators; and the SOB, or Submarine Obfuscation Body, a dummy that could lead a homing torpedo astray. There was also the MAD, a Magnetic Airborne Detector, sensitive to magnetic-field changes produced by submarines to a depth of 500 feet, allowing hunters to locate a submerged U-boat so that airplanes or surface ships could then attack it.

In October 1943 the British sent the Shell engineer Isaac Lubbock to the United States to check on progress in rocketry. He returned to London with information on two American achievements that Cherwell and Crow had been telling the War Cabinet were not feasible: liquid rocket fuel and a gas-turbine pump. If the Americans at Caltech could make these, it was reasonable to assume that the Germans could do so. But Crow still tried to dismiss photos of rockets nearing launch pads at Peenemünde as "inflated barrage balloons," until a Sandys assistant asked him why such balloons would be transported on heavy rail wagons. Crow had no answer. An October 25 Cabinet meeting featured Cherwell's attempt to dismiss Lubbock as a "third-rate engineer" meddling in matters too complex for him. In November an astute female photo-recon interpreter spotted what seemed to be ski ramps along the French coast from Cherbourg to Calais that were designed to launch some sort of rocket. But these launchers were too small for the V-2s, and Cherwell tried to use the confusion between V-1s and V-2s—the British did not yet believe that the Germans could develop *two* long-range flying weapons—to prevent further action. A timely and detailed report from the French Resistance filled in the blanks and finally made it clear that there were two separate weapons systems being developed on Peenemünde.

Then a pilotless V-1 flying bomb with a concrete warhead landed on a Danish island. It was photographed and sketched by a Danish Naval officer before being carted off by the German military. The photos were sent to Great Britain, and copies were given to the War College in the United States. Within a few days a mock-up of a V-1 was constructed in New Mexico. Proximity-fuze designers from Merle Tuve's Section T and their Canadian counterparts began to modify their product so that it would be able to bring down such a flying bomb. They were on the verge of solving the major technical problem for the proximity fuze, an internal power supply. The shells containing the fuze were going to rotate as they hurtled toward the target, so the electrical current to power the radar equipment in its nose was designed to come from a battery that made use of the centrifugal force of rotation to properly distribute the electrolyte to energize the battery.

Another Allied secret weapon that could produce large numbers of casualties, should it be unleashed against the enemy, took shape in October and November 1943, at Camp Detrick, Maryland. A "cloud chamber" was constructed to test a biological bomb and an aerosol spray that would contain anthrax, typhus, plague, or other diseases fatal to man; amounts of the toxins known to be lethal to small animals were released into the chamber and the animals' progress toward death and debilitation noted. One of the experimenters was Lord Stamp, on loan from Porton Downs. While the cloud chamber was being used to conduct tests, a four-pound bomb filled with anthrax spores, code-named "N," was manufactured from a British design in another building at Detrick. Cherwell would shortly describe the effect of this bomb in a "top secret" memo to Churchill:

> N spores may lie dormant on the ground for months or perhaps years but be raised like very fine dust by explosions, vehicles, or even people walking about . . . Half a dozen Lancasters could apparently carry enough, if spread evenly, to kill anyone found within a square mile and to render it uninhabitable thereafter. . . . This appears to be a weapon of appalling potentiality; almost more formidable, because infinitely easier to make, than Tube Alloy [the atomic bomb].

Cherwell recommended manufacture because "we cannot afford not to have N bombs in our armoury," a conclusion with which the British Chiefs of Staff agreed. Churchill asked the United States to manufacture a half million anthrax bombs. They were to be used only as a retaliatory measure if Hitler first launched something similar against an Allied country.

Ultra decrypts consistently showed that the Nazis had no intention of using poison gas or other chemical and biological warfare agents, because they believed they could not adequately protect their own people from retaliation in kind. Still, the Allies' continued fear of Nazi use of CBW was reasonable, given the character of Hitler and his willingness to exterminate millions of people. Other Allied fears of Nazi "scientific" weapons were irrational. Three weapons discussed at high levels involved a "supervirus" bacteriological weapon that would produce an incurable plague, an "endothermic" bomb that would instantly turn an area so cold that it would end all life in it, and of course the death ray, fear of which could not be allayed by assurances from scientists that such an item would continue to remain only a science-fiction nightmare.

As part of a Congress of American-Soviet Friendship in November 1943 in New York City, Soviet medical doctors talked about and showed a film of experiments in which a dead animal was restored to life by means of an "autoejector," a primitive heart-lung machine. British scientist J. B. S. Haldane had journeyed to Moscow to witness this machine in action and narrated the film about the work of a Professor Bryukonenko. The audience also saw the machine being used to bring back to life isolated organs such as a rabbit's lungs, heart, and even its detached head, which—after blood flow was restored—licked its lips in response to a drop of citric acid. Vistas of open-heart and open-brain surgery, as well as of the saving of accident and war-wound victims, were painted for the audience. In the afternoon there were discussion panels of individual sciences, led by the likes of Nobel Prize–winner in chemistry Harold Urey of Columbia and microbiologist Selman Waksman. In general, Soviet science was extolled as a model of the cooperation with governmental aims to which their counterparts in the democracies must aspire. The lesson taught by Soviet science, one participant said, "is that control and organization of science by and for the whole community does not kill the scientific spirit or initiative, nor submerge the individual scientist in a dead level of anonymity."

At Tehran in November 1943, Churchill and Roosevelt pleaded with Stalin to keep up the pressure on the Germans in Soviet territory, so that Hitler would not be able to throw his full might against the proposed Allied invasion of France. They drafted a "Declaration of German Atrocities," putting the German people on notice that at the war's end "those German officers and men and members of the Nazi

Party who have been responsible for, or have taken a consenting part in . . . atrocities, massacres and executions" would be judged and punished. Radio broadcasts, dropped leaflets, and underground newspapers soon carried the same message.

The Big Three also discussed postwar territorial moves, and Stalin promised to declare war on Japan as soon as Germany was defeated. Such optimistic palaver was premature, for the Allies continued to fear that somehow—with the V-1s and V-2s or the other secret vengeance weapons that Hitler frequently spoke about in his radio broadcasts—Germany would find ways to prevent the coming invasion, or to counter it successfully, or to make some master stroke that would once again reverse the tide of the war.

Information kept coming in steadily to the Allies regarding German research on and stockpiling of nerve gases, ten times more deadly than the mustard gas of World War I. An intercept provided the news that improved gas masks were being issued to the Wehrmacht, which raised suspicions about the Germans possessing new war gases of their own. A German scientist captured in Tunisia told his interrogators of plans to use gases against towns and fortified positions. Low concentrations of a nerve gas would cause temporary blindness and a resultant panic, without killing people. The Allies worried that Hitler might well use poison gas to block an invasion of France. There was still a worry that Japan might use gas to defend garrisons in the Pacific. After very bloody battles with Japanese forces at Guadalcanal, New Guinea, and Tarawa, American military managers themselves considered using gas against the Japanese to reduce Allied casualties in future battles.

On November 28, 1943, the merchantman *John Harvey* arrived at Bari, a southern port in Italy that had become the main supply base for Allied troops; the ship carried an unusual cargo, 2,000 large chemical bombs containing mustard gas, along with a half dozen men from the U.S. Chemical Warfare Service. Four days later 100 German bombers hit the port, causing the worst harbor disaster since Pearl Harbor—the sinking of 17 ships, including the *John Harvey*. The CWS men were killed in the explosions, and the poison gas leaked out, began to burn, and mixed with the oil on the waters to form a deadly surface film and a smoke that drifted toward the city. Since the presence of the gas aboard the ship had not been generally known, the hospitals treating the wounded did not realize that some were gas victims, and within a day hundreds of sailors—and their rescuers—had gone blind and had badly blistered and burned skin. Some recovered, some did not; nor did a thousand Italian civilians in the town, many of whom died from inhalation of the deadly smoke. An immediate cover-up was begun, with the sanction of the highest authorities; Churchill and Roosevelt agreed to full postal censorship, and the bodies of Americans were sent home accompanied by notes that the victims had died of shock or bronchitis; news of the incident was suppressed, lest the Germans use it as an excuse to start a gas war. The news got out, though, and a few months later the Allied High Command admitted there had been an accident but said the gas had been brought to Italy only for use in retaliation. As late as 1948 Eisenhower continued to dissemble about the disaster at Bari, writing in his memoir, "Fortunately the wind was offshore and the escaping gas caused no casualties."

Despite the Bari disaster, or perhaps because of it, American and Canadian experts on gas began to outline plans for G-Day, simultaneous all-out gas attacks on Germany and Japan. Thousands of tons of gas were manufactured and stockpiled, and bases for the testing of their toxicity were established in Florida and on an island off Panama. Eventually the United States would employ more than 20,000 people at a dozen locations in the research, testing, manufacture, and storage of poison gases, 10,000 of these workers at Pine Bluff, Arkansas, where they included inmates of a nearby POW camp.

By December 1943 the Allies had plenty of information that both V-1s and V-2s were being aimed at Great Britain from dozens of sites along the northern coast of the Continent. Despite Cherwell's continuing refusal to accept what Intelligence was reporting about the V-2s, he had always believed that pilotless bombs would be aimed at Great Britain, and when this was confirmed he joined the chorus urging Churchill to bomb the French coastal sites. The British bombing began, and American bombers soon participated in the raids. Sandys's Bodyline group was upgraded to "Crossbow," an Allied Intelligence committee investigation of all of Germany's long-range weaponry.

Resistance agents provided the Allies with detailed information on the locations of the sites, the thickness of their walls, the moments when the V-1s would most likely be on the ski launchers. More than a hundred permanent launching sites were destroyed, though new, more mobile, and better camouflaged (but less efficient) launchers were built to replace them. Nearly 450 Allied aircraft and 2,900 airmen lost their lives in the effort to eradicate the V-1s before they could be launched.[3] It is estimated that the bombing of the launchers set the V-1 program back six months. Eisenhower later noted the significance of this feat:

> It seemed likely that, if the German had succeeded in perfecting and using these new weapons [V-1s] six months earlier than he did, our invasion of Europe would have proven exceedingly difficult, perhaps impossible. I feel sure that if he had succeeded in using these weapons over a six-month period and particularly if he had made the Portsmouth-Southampton area one of his prime targets, Overlord might have been written off.

[3]In one of these raids, on Mimoyecques, the raiders also destroyed the most advanced launching site of the V-3, the *Hochdruckpumpe,* a huge cannon whose side chambers were to accelerate a shell out of the barrel at supersonic speed, enabling it to reach London or other targets in the British Isles before planes could be scrambled to attempt to shoot it out of the sky. Fifty V-3s were about to go into construction; none were ever fired at Great Britain.

As the Allied High Command bombed the V-1 launch sites and prepared for Operation Overlord, the invasion of Fortress Europe from the north, Colonel Boris Pash of the U.S. Army and a team of American scientists joined the Allied forces fighting their way up the Italian peninsula. Backed by General Groves of the Manhattan Project, the Pash team's mission, code name ALSOS—the Greek word for "grove"—was to seek information on whether the Germans were ahead of the Allies in the race for an atomic bomb, through investigating what the Nazis might have given to their Italian partners. They did find information touching on rocketry, guided missiles, chemical warfare, radar, and, as Pash later wrote, "on explosives in the form of personal items carried by sabotage agents," but not much having to do with atomic research, except for two scientists whom they wanted to question at length, but whom the Office of Strategic Services failed to extract from Rome.

Vannevar Bush, mindful of the leaps ahead that had come from adapting the German electric-propelled torpedo, the 75-millimeter recoilless artillery gun, and other captured weaponry, had also asked that the Pash group see what they could find in the way of weaponry that the Allied forces did not have. On February 11 Pash and his group were led to the Anzio beachhead—where a large number of Allied casualties had just been suffered—to look at an unexploded glider bomb of the sort used by the Germans to sink the battleship *Roma*, and which were then being directed against Allied troop ships. The Americans had nothing quite like the remote-controlled gliding bomb; the prize was shipped back to the United States for analysis. But the basic question of whether the Nazis were ahead in the race to produce an atomic bomb remained unanswered, at least to Groves's satisfaction. The British were already convinced that there was no German bomb, as not a single intercept decoded at Bletchley Park contained any information about such a bomb project. As with so many other puzzles, this one would have to wait for resolution until the Allies invaded Fortress Europe.

The liberation of Italy stalled when Allied forces were unable to take Cassino despite having heavily bombed the medieval monastery that was the German stronghold. One culprit was the Norden bombsight, so highly touted before the war, which was so unreliable in battle that it was useful only in daylight, and while the airplane was flying level at a constant heading and at a constant speed, and when the bombardier had the target in visual sight. New radar-controlled bombsights, though rushed to the battlefield before adequate testing, were better.

The German defenders of Cassino had been exhorted by Hitler to prevent the Allies from taking the mount or die in the attempt. While the stronghold held, Hap Arnold's office summoned Harold Edgerton of MIT to Italy. It had been two years since Edgerton had demonstrated his flash and high-speed camera for night aerial photography, but the moment and the target were now ideal. Using the two-plane system, Edgerton illuminated and photographed Cassino, pinpointing the location of German ammunition dumps and other targets for the air forces to attack, which contributed to the eventual capture of the stronghold. Edgerton promised to have even better equipment available for the Normandy invasion.

Air raids on Germany destroyed Max Planck's home and papers and the KWI for Chemistry headed by Hahn, from which Heisenberg, Mentzel, and others tried to salvage the waterlogged books of the library—but on these raids the British lost more than a hundred planes and nearly a thousand airmen. The British Cabinet was warned that the greater accuracy of German antiaircraft fire and the superior maneuverability of its fighter planes recalled from Russia and Italy were eroding the RAF's capacity to maintain the pace of bombing.

During what was known as the Big Week, at the end of February 1944, British and American air forces made five massive strikes against German aircraft-production facilities throughout Central Europe. Many factories were damaged, but perhaps the worst harm done to the Nazi cause was that the raids depleted the ranks of the

Reich's best-trained and most experienced fighter pilots, five of whom died for every Allied pilot shot down.[4]

The leadership of the Reich prepared in many ways for the inevitable Allied attempt to retake Northern Europe, among them erecting the Atlantic Wall, increasing and redistributing the production of planes and other engines of war, and speeding the manufacture of synthetic rubber and oil (at the I.G. Farben plant near Auschwitz). They had some reason for optimism, because despite the Allied bombing, German production of nearly every important implement of the war was at its height. The factories were turning out twice as many airplanes in 1944 as they had in 1940, and the production of tanks was rapid enough to outpace their losses on the Russian battlefields. V-1s and V-2s were also being manufactured; Hitler did not want to launch any of them prematurely, preferring to wait until he had enough stockpiled so that they could be fired in waves whose density and continuity would add to the vengeance weapons' power to devastate and dishearten.

Some of the thousands of V-2s that Hitler had demanded were being put together in a tunnel in the Harz Mountains, a place that became one of the most sordid of concentration camps. Prisoners from the eastern front and from France worked as slaves to complete the huge weapons: 45 feet tall, 5 feet around, with a liftoff weight of 12 tons including a payload of 1 ton. While making the missiles, about 20,000 prisoners died from starvation, disease, or outright execution. After the war Speer would try to claim that he had eased the conditions at the rocket-assemblage camp, and von Braun, Dornberger, and other of the rocketry program's leaders would try to

[4]Experience was invaluable; for instance, the ability of fighter pilots to use their new radar to raise their hit rates was totally dependent on experience. American "ace" pilots were routinely sent back to the United States at the peak of their careers, both to preserve their lives and so that their expertise could be used to train other pilots. German aces were almost never removed from combat unless injured.

insist that forcing concentration-camp prisoners to work on the V-2s was the fault of Himmler and the SS, but records show that all these men concurred in Himmler's decision.

On March 13, 1944, Wernher von Braun and two associates were drinking in a café and discussing rocketry for space travel; they were overheard, denounced to the SS, and arrested, ostensibly for speaking of rocketry in relation to something other than a military objective, but more likely because von Braun had been resistant to Himmler's plans to take over the direction of the V-2 program. Two weeks later, pressure on Hitler from Speer—then quite ill—sprang von Braun, but thereafter he and the other rocketeers continued their work under suspicion from the SS.

In March 1944, the Germans abandoned the permanent V-1 launch sites and quickly built simpler and more mobile launchers that could remain camouflaged until a few moments before the jet-powered bombs were aimed and fired. The Germans also reduced the failure rate of launched V-1s from 39 percent to below 5 percent.

To protect against the expected V-1s, the British constructed a three-tiered defense. Over and right around London was a belt of balloons. Ringing the city were antiaircraft emplacements, positioned close in to concentrate their fire and also because the V-1s were expected to fly in high and out of range until they were very near their targets. Not all the antiaircraft guns had the new American SCR-584 microwave radar-predicting device and the M-9 director for antiaircraft gunfire. Moreover, their ammunition consisted only of timed fuzes. This was partly a mistake, partly by plan. Special proximity-fuze-based shells were being manufactured for British guns, but their use over land had not yet been authorized, because General Arnold feared that if an unexploded proximity-fuze shell was captured, it could be reproduced by the Germans and used in quantity against the Allied air forces.

Another reason for not initially using the proximity-fuze-based shells against the V-1s was the presumption that the V-1s would be armed with TNT. That explosive was not vulnerable to a proximity-fuze-based shell, since it would not explode on contact but by means of the internal fuze. Thus a proximity-fuze shell would only be able to deflect a TNT-laden buzz bomb from its original target, not destroy it.

Farther from London was the third tier of the defense, a zone patrolled by fighter planes. The three-tier plan was a reasonable one, but it was not well implemented, because many of the fighter planes, antiaircraft guns, and barrage balloons were needed to protect the embarkation ports from which troops would cross to France.

In preparation for Operation Overlord, all Allied air forces had been put under Eisenhower's command. Ike ordered that direct bombing of Germany cease on April 1 and that for the next two months planes be sent against targets whose destruction would directly further the Allied invasion. Also, German defenses were taking a tremendous toll on Allied aircraft: on the night before the deadline more British airmen were killed in a massive raid on Nuremberg than had died during the entire Battle of Britain.[5]

Air Marshal Tedder and his scientific adviser, Solly Zuckerman, argued for targeting of communication and transportation facilities in France; a similar plan, used against transportation in Italy, was aiding the Allied advance there. Carl Spaatz, commander of the American air forces, preferred to target oil-storage plants and depots, whose destruction would cripple the Nazi tanks. Cherwell and Churchill wanted the airpower directed only at areas in which ground fighting would take place, because to do otherwise would surely result in the deaths of many French civilians in the vicinity of

[5]American air raids, conducted during daylight hours, lost fewer bombers because after December 1943 they were protected by the American-made Mustang P-51B fighters, which were better able than their British equivalents to counter German fighters.

the communications and transport facilities. The dispute was not resolved until both President Roosevelt and General Charles de Gaulle agreed with Zuckerman and Tedder. Eisenhower, in an attempt to include everyone's objectives, directed the bombing to be done against prospective landing sites, communications nexuses, and rail lines, with any "strays" to have the highest chance of hitting oil depots. In April and May, with the aid of the Resistance forces who helped to precisely locate the targets, Nazi communications centers, transport lines, and oil depots were relentlessly bombed, with lower loss of life on the ground than had been expected. An additional dividend for the Allies was that during the raids its planes were able to destroy more than 5,000 German fighter planes. The combination brought about Allied air superiority over Northwestern Europe—on D-Day the Luftwaffe would be able to fly only 319 sorties over France, while there were 10,500 Allied sorties—and the reduction of the German supply lines, bombed down from the 100 trains per day required to keep the German Army functioning properly to thirty-two per day. Air raids also diminished the German surface Navy, one raid on Le Havre damaging thirty-eight of its large ships.

During the months of preparation for the invasion, Frédéric Joliot, as president, and the other members of the Directing Committee of the National Front in France met frequently to decide on matters of importance to the Resistance, to coordinate activities of various units, sometimes to hear secret emissaries from London, and occasionally to agree to the execution of a traitor; in such latter instances, Spencer Weart writes, "the final decision was said to have been Joliot's." Joliot's former subdirector of the Collège de France laboratory had recently become chief of the laboratory for the Prefecture of Police and was regularly passing on information about forthcoming Nazi raids on Resistance cells. It was likely to have been a tip from that source that spurred Joliot to organize an escape to Switzerland for Langevin in early May. While Irène Joliot-Curie lived with their children at Arcouest, Frédéric remained in Paris. He went into hiding under a false name and an identity as an electrical engineer.

As D-Day approached, the Resistance was readied for widespread organized sabotage and open combat against the German occupiers of France; arms, communications equipment, and detailed plans were dropped by parachute or smuggled in, so that the actions of the Resistance could be coordinated with and assist the Allied battle tactics.

Meanwhile through the early spring of 1944, the fortunes of the Allies gained on other fronts. Soviet armies in three areas succeeded in taking back a great deal of territory that had been under German control and in capturing tens of thousands of German soldiers. Hitler's need to station his armies in the east, to prevent Stalin's troops from overrunning German territory, seriously depleted the resources of personnel and matériel that the Nazis had available for use in Northern Europe against the expected Allied invasion.

In India indigenous and British troops succeeded in halting the advance of the Japanese land forces, which were also kept busy fighting off newly invigorated Chinese armies. In the South Pacific, American, Australian, and other Allied forces steadily took dozens of islands and strongholds from the Japanese, notably Hollandia in New Guinea. In nearly every instance in which Allied forces wrested an island or a base from its defenders, Japanese casualties were phenomenally high—between 80 and 95 percent of all troops—because many committed suicide rather than be captured; often the only Japanese to surrender were those too ill or injured to kill themselves.

On the high seas American naval forces in the Pacific used every bit of advanced scientific equipment and weaponry available, including the antiaircraft shells with proximity fuzes, to gain control of vast areas of the ocean; Admiral King, once converted to the idea that scientists could help the Navy accomplish its tasks, became an enthusiast for radar, Hedgehogs, proximity fuzes, and OR analysis.

The march of the Allies ever closer to the Japanese home islands was aided by a recent code breakthrough, itself the result of the capture of Japanese Army codebooks and enciphering devices. Intercepts of coded information enabled MacArthur and other commanders to know the precise size and location of the defenders' deployments—in several places it allowed unopposed landing of Allied troops; as a secret official report from U.S. Intelligence later put it, "Never has a commander gone into battle as did the Allied Commander Southwest Pacific, knowing so much about the enemy."

American submarines, equipped with better torpedoes, radar, and with the benefit of intelligence about ship positions obtained from decryptions, decimated Japanese shipping. Adapting tactics from the wolf packs, American submariners had their own "happy time," able to easily pick off Japanese merchantmen, troopships, and oil tankers. But while the U-boats' attempt to cut off supplies to Great Britain had failed, American submarines were succeeding in 1944 in slowly strangling Japan's home islands. At the same time Allied destroyers and shore-based planes sank dozens of Japanese subs, many of them by means of Hedgehog depth-charge launchers.

In May 1944 it occurred to SS leader Himmler that among the Jews now arriving at various concentration camps, "there are doubtlessly quite a lot of physicists, chemists and other scientists" whose talents could be utilized by the Reich. Himmler ordered the establishment of an institute in a concentration camp where "the expert knowledge of these people can be set to work for the man-power and time-consuming tasks of calculation of formulae, finishing of specific constructions, but also for basic research." Concentration camps were scoured for likely candidates, calculators were imported from the subjugated countries, and several dozen prisoners were set to work in Sachsenhausen—some of their names supplied by former colleagues, in an attempt to rescue at least these Jewish scientists from certain death.

The temporary reprieve of these Jewish scientists was a minor counterpoint to the increasing intensity and reach of the Nazis' extermination program. As the likelihood of an Allied invasion rose, so did the Nazis' efforts to round up Jews from every corner of the Continent—from Hungary, where they had been protected by the puppet regime for two years, from Vichy France, and from the eastern front. The ovens at Birkenau had to work at full capacity to reduce to ashes the bodies of so many executed prisoners.

After a very accurate forecast from Allied meteorologists that showed a window of good weather during a period of bad weather—at a moment when decoded intercepts showed that German forecasters anticipated an unbroken stretch of bad weather—the Allied invasion of Fortress Europe began, on the night of June 5–6, 1944. The landing purposely took place on a night when the moon would rise at between one and two in the morning, aiding the early parachutists and deceptions; on a morning when half tide would occur forty minutes after daybreak; and when the days of the year were at their longest, which would curtail the Germans' ability to move around troops and resources at night.

Operation Overlord was the largest single military action ever mounted; some 5,000 ships, 11,000 Allied planes, 150,000 troops disembarked on the first day. The possibility of such a massive invasion had been discounted by Hitler and his generals, though it had been clear since the Americans entered the war thirty months earlier that the full industrial capacity of the United States, Canada, the Commonwealth countries, and the overseas outposts of France and the Netherlands would be brought to bear on the Third Reich. In that sense the invasion of Fortress Europe was retribution for the Nazis' willful ignorance of rationality and the scientific method: any competent statistician with access to published figures could have told Hitler that such an invasion was coming and what its strength would be.

So much that was new in the history of warfare went into the Allied invasion and the German attempts to repel the invasion that it is difficult to isolate and identify those elements to which scientists contributed most significantly. To name just a few on the Allied side, the troops carried with them new gas masks with small canisters attached to permit breathing of contaminated air; medic units carried whole blood; never-before-used weapons such as antitank explosive packets were issued to some units; and landings were preceded by barrages of rockets launched from seagoing barges; there were flamethrowing tanks, amphibious tanks, and tanks without turrets whose only function was to lift other tanks across dikes. Among Churchill's better science-based ideas was to urge the construction of two artificial harbors off Normandy, called "Mulberries," at which ships could anchor to unload. He also insisted that twenty pipelines be built under the Channel so that fuel oil could flow from Great Britain to France for use by tanks and other vehicles once ashore—the project was dubbed PLUTO, Pipe Line Under The Ocean.

Many new weapons, transport, and communications technologies developed thus far during the war by the Allies were used in Overlord. In several instances the weaponry had been deliberately held back until this moment so that the Germans would have no prior knowledge of it and would be unable to counter it. Almost the only technology not well enough developed by the Allies to see battle in the first weeks of Operation Overlord was the jet plane; the RAF Meteor 1 jets would not take part in combat until July and August.

German basic fighting equipment had been technologically enhanced and, weapon for weapon, was superior to that of their Allied counterparts. German gun barrels lasted far longer than did those of the Allies, German antitank weapons were better than those of the Allies, German "sniper-scopes" had no Allied parallel. The R4/M rocket projectiles carried by the FW 190 were judged by British battlefield-recovery teams to be faster and more powerful than any British rockets, and the small and portable MX 108 30-millimeter antiaircraft gun was considered "unique" and very effective against Allied aircraft.

Where the Allies were clearly superior was in the full application of all sorts of scientific disciplines to the invasion preparation and landings. Geologists, soil analysts, experts on tides, and Operations Research analysts were involved in many phases. Psychologists as well as magicians participated in the elaborate deceptions designed to mislead Germany about the timing and location of the invasion; the Germans became convinced of the existence of the fictitious First United States Army Group and Twelfth British Army, and of supposedly major campaigns aimed at Norway, Sweden, Turkey, Greece, Marseille, and other far-flung destinations.

Planning of the movement of millions of men and multiple tons of material to the Continent required the services of every piece of advanced computing machinery available in the United States and Great Britain, such as the Rockefeller differential analyzer at MIT. Interception and analysis of German communications, an essential contributor to the invasion's success, was done in part by the established Bombes that decrypted Enigma-based communications and in part by the recently developed Colossus machine that decrypted teleprinter transmissions, the usual way in which Hitler and his generals communicated. The Colossus used vacuum tubes—then called theriotic valves—to do the jobs that Turing's relays had done in the Bombes; since the vacuum tubes had no moving parts, they were a thousand times faster than the electrical relays, and so the Colossus could process 5,000 frames of teleprinter tape per second, which enabled the initial analysis of the average message, with its billions of possible combinations of code wheels and letters, to be performed in half an hour. After the Colossus completed the initial analysis, the coded message could then be more readily decyphered by other means. Ultra, the combined product of all the decryptions of German messages, played a major role in Operation Overlord by giving Allied planners vital information on German orders of battle, deployments, depths of supplies, localized weather conditions, and dozens of other factors. In the end the great accomplishment of the Normandy invasion was not only the success of the well-trained armies and the triumph of logistics, it was also the victory of the scientific method applied to warfare.

One week after D-Day, while the Allied invasion of Normandy was proceeding much as planned, the first V-1s entered British airspace. One exploded at Swanscombe, where it damaged buildings, and another at Bethnal, where it killed six people. The immediate reaction in the War Cabinet was relief: the dreaded buzz bombs were not filled with poison gas. Once more the nightmare of gas dropping from the skies had not materialized.

Nonetheless, explosive death was raining from the skies, and it seemed to the public almost as bad as gas. In the days that followed, more than 9,000 V-1s were launched at Great Britain: each day and night there were dozens, then hundreds of them crossing the sky and causing confusion and terror in London; in the "Bomb Alley" of Kent, Sussex, and Surrey; and in the embarkation ports of Portsmouth and Southampton. That these fiery bombs were *Vergeltungswaffen*, "vengeance weapons," became immediately obvious, because in addition to physical damage they also produced psychological harm, clusters of them assaulting the populace, often in cloudy weather or during the darkest hours of the night. The V-1s' screaming mixed with the whines of fighter planes trying to shoot them down, with the staccato punctuation of antiaircraft guns, and with the wail of sirens from fire engines speeding to put out the conflagrations. A robot bomb was like something out of a science-fiction horror story, screaming and blazing across the sky until the jet motor shut down and there was a stunning silence, the more deafening because it presaged the bomb's imminent explosion below on earth.

For a week the British public lived in absolute fear of the flying bombs because the government refused to tell that it knew what these abominable things were and had been waiting for their attack. In reaction to the strange noises emanating from the V-1s, Londoners labeled them "Doodlebugs," a nickname that helped to balance their terror. The bomb was purported by the press to be able to flatten an entire London block or a country parish. That was stretching things, but only a bit, since the V-1 did carry a payload of 850 kilograms, and a direct hit by a V-1 could damage a circle of adjacent buildings, no matter how sturdy their brick or steel.

The British air defenses fought against the V-1s, but because the bombs were coming in low—lower than expected, at about 2,300 feet—the antiaircraft guns had less time available to locate and try to shoot them down before they landed.

The buzz bombs were eight times more difficult for a fighter plane to bring down than a German bomber flying a straight course. For a plane to shoot down a V-1 was also dangerous, as more than one exploded bomb took the attacking plane with it to a crash landing. Some pilots nudged the V-1s off course with their planes' wingtips when bullets did not suffice; others deflected the bombs by means of their slip-streams.

In the third week the British government ordered the evacuation of all nonessential personnel from London, more than a million people. When this news reached Germany, Goebbels took to his own airwaves, crowing that "London is chaotic with panic and terror. The roads are choked with fleeing refugees." This thought heartened the German public, which had suffered through the "dehousing" bombings as well as a string of defeats from El Alamein to Stalingrad to the loss of the entire Italian peninsula, and which was beginning to learn the dimensions the Allied invasion of Normandy.

In the midst of the V-1 assault on Great Britain's cities, Prime Minister Winston Churchill wrote to the British Chiefs of Staff demanding that very serious thought be given "over this question of using poison gas." He had recently received an assessment from Porton Downs, the gas development and testing facility, which indicated that they had enough bombs to paralyze the cities of Hamburg and Berlin and even to kill all the inhabitants of those cities—on a single night. Churchill, after warning that gas could not be used unless it were a life-and-death matter or unless "it would shorten the war by a year," asked for a "cold-blooded calculation . . . as to how it would pay us to use poison gas, by which I mean principally mustard. . . . I want the matter studied in cold blood by sensible people and not by that particular set of psalm-singing uniformed defeatists which one runs across now here now there."

Churchill believed that Allied air superiority would prevent German bombers from getting through to retaliate, and that the Allies had more gas than the Nazis and could use it to hurt the Germans so badly that their response would be weak at best. The prime minister did not then know about or understand the toxicity of the German nerve gases, of which Hitler had 12,000 tons already loaded into bombs and shells.

The British Chiefs of Staff refused to go along with the prime minister's request, acknowledging that while it was "true" that the Allies could "drench the big German cities with an immeasurably greater weight of gas than the Germans could put down in this country," the Germans could indeed retaliate against British civilians, whose morale would suffer. Churchill grumbled, "I am not at all convinced by this negative report. But clearly I cannot make head against the parsons and the warriors at the same time."

The second waves of invading Allied troops in Normandy carried additional scientific equipment to assist in the battles—for instance, the British Gee system radars for directing bombers. These were set up in France to provide greater guidance for strikes that went farther east into Germany than those that had been launched from Great Britain.

General George S. Patton's tanks were pushing so quickly through France toward the Loire Valley that they were outrunning their information on German troop and tank movements. Harold Edgerton of MIT was asked to use his strobe and flash camera system to photograph German lines at night; the results permitted Patton to route his tank columns around the German Panzers and to maintain the speed of his advance. Edgerton's work was so important to the Allies' push that the MIT professor would remain in France through the summer, the fall, and into the winter of 1944.

Nearly simultaneous with the Allied invasion of Fortress Europe in June, Soviet troops began an offensive in the east, with three armies in three locations, that would eventually cost Germany some 25 divisions. Soviet forces advanced across Poland and through Bessarabia, as well as south toward the Romanian oilfields. The continuing pressure from the Soviets also prevented Hitler from allocating more forces in the west to try to counter the Allied thrust. Also at the same time, in the Pacific, Allied forces began a multifaceted assault on the Marianas, a group of islands almost a thousand miles nearer to Japan than are the Solomons and New Guinea. It began with the largest air strike of the Pacific War, 200 bombers directed against various airfields, followed by a massive invasion of Saipan. The Japanese high command was caught off guard by these attacks, having presumed that the Allies would next target the Palau Islands, closer to the Philippines. "Hell is on us," one of the emperor's advisers is said to have observed when the news of the Saipan attack reached Tokyo. The Japanese navy attempted to come to the aid of the defenders of Saipan in the Battle of the Philippine Sea, which only resulted in the near-annihilation of the Japanese carrier-based airplanes and of the bulk of its ships. The Allied recapture of the former American base island of Guam was more difficult but was eventually accomplished, along with the subduing of the rest of the Marianas; by late August the Allies were threatening the Palaus and also preparing to return to the Philippines. As a result of the defeats,

Admiral Tojo and his cabinet resigned on July 18, 1944, and were replaced by less experienced generals and ministers.

On July 20, 1944, a plot by German military officials to assassinate Hitler failed. The professional military leadership of the Reich, some of whom had been involved in the plot, were relieved of command. A Total Mobilization of Resources for War was ordered, under the propaganda minister, Goebbels, and Himmler was given command over production of the V-1 and V-2 rockets. Speer remained in place, primarily because that month armaments production in the Reich reached its peak, but he saw "the first clear signs of disintegration," Nazi Party loyalists ignoring Hitler's pronouncements, good military men suspected of treason, the ordering of the destruction of industrial plants in the path of the Allied and Soviet advances.

In July 1944, to counter the V-1s coming at London, many barrage balloons were brought back from the Channel ports to the capital. That enlarged the barrier from 500 to 2,000 balloons. Some 500 searchlights were set up with 20,000 troops assigned to attend them. Meanwhile the antiaircraft guns went the other way: in a hurried two days, the entire phalanx of 1,100 guns was moved from London to the Channel coast. There they could fire at incoming buzz bombs over expanses of water and sparsely populated land—and without fear of their proximity-fuze shells hitting Allied fighters, which were repositioned closer to London.

During the early days of the V-1 attack, Merle Tuve of Section T had not understood why the British were not responding to his requests for information on how the proximity fuzes were working and were not sending rush orders for more. E. O. Salant of New York University, sent to Great Britain, soon found the reason for the silence: General Frederick Pile, commander of the British air defense forces, had never received the information that tens of thousands of proximity-fuze shells had been manufactured for his antiaircraft guns; the War Office had simply filed the fact and never passed it along to Pile. Proximity-fuze shells were quickly sent over to Great Britain and put into operation. And they worked well, in part because the Germans, instead of using TNT in the V-1s (as the Allies had expected), used a more sensitive explosive, one that could be set off if a splinter struck it—which was precisely what the proximity fuze was designed to make happen.

But even when the proximity-fuze shells were available, many antiaircraft crews did not use them. Pile told Salant that his men (and women, since nearly half the crews were women) claimed that the proximity-fuze shells were going off against clouds. Operations Research had stations alongside the antiaircraft batteries and worked with Salant to examine the records, which showed that as many of the shells were exploding prematurely in clear weather as in cloudy conditions. This led to a directive to use the fuzes in any weather. But the battery crews had also discovered that the wax coating on the shells made a good chewing gum; and, of course, scraping the wax off compromised the shells, so a further directive had to be issued: leave the wax coating alone.

The more serious problem was that the rapid flight of the V-1s—400 miles an hour—made it difficult for a battery crew to keep up with a buzz bomb's transit and fire shells at it as it approached. C. Stark Draper from MIT, the former stunt pilot and designer of stability systems for planes and projectiles, came over to investigate. On the spot, Draper modified the M-9 fire director so that the antiaircraft-gun crews were better able to follow the V-1s and to direct the proximity-fuze shells at them.

At first the Allies were perplexed that the Germans did not seem to realize that proximity fuzes were being used, because it was so easy to figure out: timed-fuze shells exploded in a scattered pattern in the air; proximity-fuze shells exploded in clusters around the target object—and that pattern was visible from as far away as the French coast. The Germans' inability or refusal to recognize that a proximity fuze was being used was fortunate for the Allies, since if they had realized, they could have quickly modified the V-1s—used TNT rather than the more sensitive explosive, for instance. Later on, the Allies learned that the Germans had indeed noted the unusual pattern of the explosions but never bothered asking their experts to explain what might have caused it. Another odd aspect of the V-1 was that it, like the German electric torpedo, had been designed too well—its flight pattern was so straight and true that it could be mapped and hit; had the V-1s deviated slightly from their appointed path or wandered a bit from the mean as they traveled—as most other projectiles did—the new American firedirectors and proximity-fuze shells would not have been as effective against them.

In the first months of the concentrated launchings of the V-1s, 5,000 people were killed and another 16,000 injured, more than during the shank of the 1940–41 Blitz; eventually the V-1s would kill and injure about 50,000 Britons and destroy a million residences, 149 schools, 111 churches, 98 hospitals, and several royal palaces. Most of the damage and the deaths occurred in the first fifty days of the launches, before the complete trio of new American-made weapons were finally all put in place and the antiaircraft crews were comfortable with them.

During those first fifty days of V-1 assaults, most of the buzz bombs did hit Great Britain, and the remainder were destroyed by planes or malfunctions. After the adjustments of Salant and Draper, and when the proximity-fuze shells and firedirectors had been put into full operation, in the first week of use this trio of innovations helped antiaircraft fire destroy 24 percent of the V-1s they engaged; the following week the percentage was 46, the third week it reached 67, and in the fourth week it was 79 percent.

On the last day on which a large phalanx of the bombs was launched at Great Britain, early-warning radar picked up 104; sixteen malfunctioned and ditched in the English Channel, sixteen more either were shot down by fighter planes or became entangled in barrage-balloon cables or nets and did not explode near their targets, and sixty-eight were destroyed by antiaircraft fire. Only four reached London. By that time, late August 1944, the Allies had captured and put out of commission the launcher sites in Cherbourg and the Pas de Calais.

General Pile would shortly send to his superiors a report on repulsing the V-1 attacks, forwarding a copy to Vannevar Bush, "With my compliments to OSRD who made the victory possible." An effusive letter to General Marshall conveyed the same sentiments.

On September 7, 1944, Duncan Sandys reported to a packed press conference in London that the nightmare of the buzz bombs was over. His timing was terrible. On the very next day a sonic boom sounded over London, startling everyone: it came from the first of more than a thousand V-2 rockets that would shortly be launched at the British capital.

| CHAPTER ELEVEN |

Ending the War

As the Allied and Free French forces neared Paris on August 19, 1944, Frédéric Joliot, who had been living in the city as Jean-Pierre Caumont, an electrical engineer, in order to further the Resistance, packed two valises full of acid and potassium chlorate from his laboratory at the Collège de France and took them to the prefecture of police, whose laboratory was run by his former subordinate. Joliot had been waiting for this moment for months; just before the invasion he had spirited his family over the border into Switzerland and had addressed a statement to the Collège de France faculty under the title "We shall no longer yield." It spoke of the valiant stances taken by universities in other occupied countries, as opposed to that of the universities in Paris, which "supported oppression without moving," and it evoked the names of science colleagues who had been arrested and of the scientists who had been deported to work in Germany. "Do you not feel the passion of revolt rise within you when you know that the young men, who revive through your teaching the best of yourself so that France may live, are claimed by Hitler to work in Germany against France. . . . We can no longer remain inactive in the face of systematic destruction of everything which provides our reason for living." Now Joliot was going to act on his own words. With his friend he descended into the wine cellar of the police headquarters, where they located and emptied bottles of champagne and refilled them with acid and gasoline, then wrapped them with paper soaked in potassium chlorate to make Molotov cocktails. Next day Joliot directed three laboratories at as many institutions in making homemade bombs that were used in street battles against German tanks.

The Free French forces entered Paris on August 25, and two days later the ALSOS team, with a clutch of American scientists under the direction of Boris Pash, reached the city with orders to find and interrogate Joliot. A dozen such teams had been sent into the field by the American, British, and Soviet armed and secret services. On the Pash team was Albert Noyes, professor of chemistry at the University of Rochester and the leader of an OSRD research division. Noyes had known Joliot since their wives had been student colleagues together. Another was Samuel Goudsmit, the Dutch-born Jewish nuclear physicist; he had been working most recently at the Rad Lab at MIT and had been tapped by General Groves to lead the scientific part of the ALSOS team, specifically to find out how far along the Germans were in making an atomic bomb.[1] Goudsmit was well prepared for his ALSOS mission; he spoke several languages and was also a nuclear physicist of repute; Isaac Rabi would later avow that the early work of Goudsmit in association with George Uhlenbeck "was a tremendous feat. Why these two men never received a Nobel Prize for it will always be a mystery to me."

The ALSOS team had first looked for Joliot at Arcouest and had

[1]Goudsmit was chosen in part because although he was a nuclear physicist he was not working with the Manhattan Project, and if captured he could not reveal Allied atomic secrets in any useful detail.

to fight their way to his summer home, only to learn that he was not there. In Paris they were told that Joliot was in hiding, but Pash picked up the telephone, dialed Joliot's laboratory at the Collège de France, got him on the line, and told him they were coming over. "He welcomed us warmly," Noyes later recalled. "There was little to eat, we had only our K rations, but that first night we had a real banquet in the laboratory. As we did not know where to go, we slept on the spot in camp beds." They wanted to know about nuclear research, and Joliot had little to tell them. Shortly Pash tricked Joliot into agreeing to go to London, ostensibly to meet his former colleagues but actually to be more extensively debriefed.

Goudsmit pursued one lead that took him to the Netherlands, and he went to visit his boyhood home in The Hague, from which his parents had been sent to a concentration camp in 1943. The house was a shambles, a shell whose wooden stairs and beams had been purloined by neighbors to use as fuel during a cold winter. "I wept for the heavy feeling of guilt in me," he later wrote; he believed that had he acted more quickly on obtaining visas and passage to the United States for his parents in May 1940, "surely I could have rescued them from the Nazis in time."

Pash, Goudsmit, and their team eventually found what the Nazis had hidden: the Phillips plant at Einthoven, a major treasure trove of technologically advanced machinery and, alongside the Albert canal, 70 tons of uranium ore. That ore was shipped to the United States, along with another 30 tons—four train-car loads—found in the Bordeaux region of France; in the United States the ore would be refined and used to produce the critical mass in the atomic bomb to be dropped on Hiroshima.

From September 8 on, the V-2s kept coming at Great Britain. Planes, balloons, and antiaircraft guns were no defense against a missile that took only about five minutes to travel from the launch sites in Holland to London, that descended toward a target from a height of several dozen miles at supersonic speeds. Radar might warn of the attack, but the V-2 would come so swiftly after the warning that it was seldom possible to reach an air-raid shelter for safety. When a missile hit, the devastation was severe, twice that caused by a V-1. Allied authorities were thankful that the Germans had not figured out how to install a proximity-type fuze in the V-2, which would have enabled it to do even greater damage. Although the terror of the V-2s was worst against civilian-occupied targets—160 shoppers were killed in a direct hit on a Woolworth's department store—their effectiveness was highest when they hit rail lines, as 358 of them did, nearly succeeding in paralyzing British rail transport. Of the 1,000 V-2s launched at London, 500 would hit the city, causing massive damage. But overall, as Speer would later comment,

> The whole notion [of the V-2s as a decisive weapon] was absurd. The fleets of enemy bombers in 1944 were dropping an average of 3,000 tons of bombs a day over a span of several months. And Hitler wanted to retaliate with 30 rockets that would have carried 24 tons of explosives to England daily. This was equivalent to the bomb load of only 12 Flying Fortresses.

In retrospect, Speer thought ill conceived the diversion of resources necessary to produce the V-2 away from the production of conventional weapons and of those new science-based weapons that could have had more impact on the Allies' ability to wage war—like the Waterfall antiaircraft rocket, which might have curtailed Allied air raids against Germany, or jet fighters, which might have prevented the Allies from achieving air superiority over Normandy. Yet "I not only went along with this decision on Hitler's part but also supported it . . . Our most expensive project was also our most foolish one. Those rockets, which were our pride and for a time my favorite armaments project, proved to be nothing but a mistaken investment . . . an almost total failure."

Germany's enemies coveted the technologies—and the scientists—involved in the creation of the V-2s. The desire to acquire the missiles and their creators affected the directions pursued by the Allied military units and the territory each would control at the war's end. While the English-speaking forces were occupied in pressing south in France and east toward Germany, Soviet forces moved through Poland, intent on reaching such sites as Blizna, where German missiles had been tested. Churchill hurriedly sent Stalin a cable asking permission for Allied observers to go into that site when it was captured. In mid-August, a few Allied observers and experts arrived near Blizna and were told by the Soviets that the town had not yet been captured—even though it was obvious that the Soviets had taken control of the area. The observers were kept out for several more weeks, then allowed to take only debris from the site, the rockets themselves having already been removed; upon returning home, the observers found that what they had carried back consisted of rusting aircraft parts.

Because of the Soviet push through Poland toward Germany, in October 1944 Himmler ordered the destruction of the death-camp machinery at Auschwitz, so that there would be no evidence of the mass killings to be found and shown to the rest of the world.

By the fall of 1944 "factories" had been set up in a few Japanese schoolyards for the manufacture of balloons that could convey bombs on the jet stream to the North American continent. The girls worked night and day, sometimes given drugs so they could stay alert and ignore their hunger, making sheets of paper-and-paste or of rubberized paper; military workers then filled the balloons with hydrogen gas and attached antipersonnel and incendiary bombs. More than 9,000 of these balloon bombs were launched; most sank in the Pacific Ocean, but some did reach Canada and the United States, started forest fires, and in a few instances killed people—for instance, five children and a woman having a picnic in Oregon. American and Canadian authorities were very alarmed by these balloons, because they could well have contained biological-warfare agents rather than ordinary explosives; every balloon that landed and could be located was examined by medical, agricultural, and Chemical Warfare Service agents, and in some cases the remains were immediately frozen and sent to laboratories for analysis. Canada readied a planeful of peat, to be infected with bubonic plague, for retaliation if any of the Japanese balloons were determined to have carried diseases dangerous to animals or human beings.

More serious desperate measures were begun by Japan in mid-October of 1944, with the instigation of the "Divine Wind," the kamikaze squadrons of young pilots who had agreed to commit suicide. "The only way for assuring our meager strength will be effective to a maximum degree," wrote the Japanese admiral who assumed command of the air fleet in Manila, "is for our bomb-laden fighter planes to crash-dive into the decks of the enemy carriers." In one sense Japanese belief in the power of the kamikazes—in human willpower—to reverse the tide of the war was the opposite of the belief that Hitler would unleash a secret, scientifically advanced and technologically superior weapon to win the war at the last moment; but in another sense the Japanese and German beliefs were identical: a wish to magically pluck victory from the jaws of impending defeat.

The kamikazes were instruments of terror; the death toll from them mounted steadily, and their attacks disheartened sailors aboard the targeted ships. But while they did disable and sink dozens of Allied ships in the battles taking place as the Allied forces moved ever closer to the Japanese home islands, they did not materially set back the Allied timetable. The suicide squadrons would have been more deadly had the Allies not been using proximity-fuze antiaircraft shells, which brought down or deflected aside many kamikazes before they could crash into ships.

To repel Allied air attacks during the Battle of Leyte Gulf, the Japanese superbattleships *Musashi* and *Yamato*—believed unsinkable by their crews, with 400-millimeter-thick armor plate—utilized the largest naval guns in the world, 18-inchers, to fire experimental "Sanshikidon" fragmentation bombs at the enemy aircraft. The use of this new ammunition all but wrecked the barrels of the ships' big guns, rendering them useless for firing regular shells. Moreover, the Sanshikidons did not stop Allied torpedo planes from sinking the *Musashi* and badly damaging the *Yamato*. Similarly, other new Japanese ammunition, steel-piercing shells designed only to explode upon hitting reinforced armor plate, were stymied in action against small American "jeep" carriers. Because those carriers' external plates were thin, the Japanese shells passed right through them and did not explode.

In Germany during the fall of 1944, Dr. Robert Ley, Hitler's labor commissioner and a chemist by trade, touted to Speer the wonders of "this new poison gas" tabun and urged that the time had come to use it against Soviet troops pressing in on Nazi territory at several points. Goebbels, Himmler, and other top Nazis agreed with Ley and broached the subject with Hitler at a situation conference. According to Speer, Hitler "hinted" that use of the gas against the Soviets might be acceptable to the British and the Americans, whose governments "had an interest in stopping the Russian advance." None of the military chieftains at the conference backed the use of tabun or sarin just then, and "Hitler did not return to the subject."

Fighting on Germany's western borders continued through the fall of 1944. Still based in France, the American ALSOS team led by Pash and Goudsmit was eager to get to Strasbourg; a plethora of clues and signs from Paris, Belgium, and the Netherlands all suggested that the university at Strasbourg had a nuclear-physics program. The city was in contention until November 15, and after the ALSOS team entered it they could not find the nuclear program, because as Goudsmit later wrote, "the nuclear laboratory was occupying a wing of the famous Strasbourg Hospital and the German physicists were trying to pass as medics." They found some scientists, but not von Weizsäcker. However, his papers were there, and Goudsmit sat down to read them by candlelight and gas lamp. "Here, in apparently harmless communications, was hidden a wealth of secret information available to anyone who understood it." The haul included "ordinary memos," notes about the difficulty of obtaining uranium, notes of computations, shorthand notes of secret meetings on nuclear matters, and a copy of a letter to *"Lieber Werner,"* with Heisenberg's address and phone number to which the Berlin nuclear work had been transferred. While "no precise information was given . . . there was far more than enough to get a view of the whole uranium project." The papers showed

> that the Germans had been unsuccessful in their attempts to separate U-235. They had probably started separation on a very small scale by means of a centrifuge and they were working hard to construct a uranium pile . . . [but their] computation showed that as of August, 1944, their pile work was still in a very early state. They had not yet succeeded in producing a chain reaction.

After two days of study, Goudsmit thought he had the answer to everyone's question: were the Germans on the verge of completing a nuclear bomb? With the documents, several scientist prisoners, and that answer, he and Pash headed to Paris to meet Vannevar Bush.

In the United States, Bush had become a celebrity, immortalized on the cover of *Time* magazine as the "general of physics" leading an agency that was "regarded almost with awe" by Washington insiders. He had recently decided that once the war was over the OSRD ought to be disbanded, but that the notion of the government's funding contract research by the universities and industrial companies should survive. He drafted a letter for President Roosevelt's signature, addressed to himself, asking that Bush prepare a report on how OSRD's experience should be used "in the days of peace ahead" for purposes such as the improvement of health, the raising of the standard of living, and the creation of new jobs. "New frontiers of the mind are before us," the Bush-drafted letter said, and posed four questions, the most important being "What can the government do now and in the future to aid research activities by public and private organizations?" Bush's response would be "Science—The Endless Frontier," a manifesto urging the creation of a National Research Foundation, the setting of long-term military R&D goals, and permanent federal support of research by outside institutions for the benefit of government departments.

In the fall of 1944, while the letter and the response were still in formation, Bush determined to use some of his clout in Washington to fight the Joint Chiefs' ban on the use of the proximity fuze over land. Here was an odd turnabout: the civilian scientist chiding the military because it would not use a science-based terror weapon even after that weapon had proved its worth in battle. The military's basic argument against using it over land had been that should one be captured unexploded, it might give the enemy a chance to reproduce proximity-fuze shells and use them against the Allies. Obtaining opinions from engineers that it would take Japan or Germany two years to produce such fuzes in quantity, Bush tried to convince his old nemesis, Admiral Ernest King, that howitzers firing proximity-fuze shells could be a devastating antipersonnel weapon. "I have agreed to meet with you," King said, "but this is a military question, and it must be decided on a military basis, to which you can hardly contribute." Bush retorted that the proximity fuze was "a combined military and technical question, and on the latter you are a babe in arms and not entitled to an opinion."

On the day of that argument, October 24, 1944, V-1s began appearing in the skies over Antwerp. Field Marshal Bernard Law Montgomery's forces had recently liberated the Dutch port. Its thirty miles of wharves, facilities for storing 100 million gallons of oil, and associated machinery made it perfect for supporting the Allied armies' advance into Germany, which was why the Germans sent V-1s and V-2s to destroy Antwerp. News of the October 25 attack so alarmed the Allied Chiefs of Staff that the next day they authorized for the first time the use of the proximity fuze over land, for whenever the V-1s should start falling in quantity. There was nothing that could be done to counter the V-2s.

The following day Bush set out for France. At Versailles, General Bedell Smith agreed with Bush that proximity-fuze shells for howitzers ought to be gotten into the hands of troops. Smith asked whether the Germans would be able to make an atomic bomb that could impede the Allied liberation of Europe. Bush did not answer the question just then. He spent a week talking to local ordnance officers near the front lines, explaining the use of the proximity fuze to them—and then met with Pash and Goudsmit. Sam Goudsmit advised Bush that the ALSOS interrogations of German and French scientists so far, and the evidence they had found, brought him to the "unmistakable" conclusion "that Germany had no atomic bomb and was not likely to have one in any reasonable time. There was no need to fear any attack from an atomic explosive or from radioactive poisons."

Now Bush, on a return visit to Bedell Smith, could tell Eisenhower's aide that there was no need to hasten the Allied timetable for defeating Germany in order to short-circuit a possible German atomic bomb. Bush's report was one reason that American advance did not make a faster push to Berlin.

But General Groves was not wholly convinced that there was no remaining threat from the German nuclear program, and he directed the ALSOS team to keep searching—for the uranium pile, for the uranium that had been taken from Belgium, and for the key scientists, the most important being Werner Heisenberg. They were to do so, as Pash later recalled, "in order to end the German atomic effort once and for all—and also in order that they should not fall into Soviet hands!"

So confident were the Allies that the war in Europe would soon be over that in late November 1944 a press conference was held in New York City at which the greatest American scientific secret of the prewar period—the one sought in vain by Great Britain, France, and even the USSR—was made public. A Norden bombsight went on display at a Rockefeller Center museum, replete with a mock-up of the cabin of a Navy Liberator from which an approach-and-bombing run over Germany could be simulated. One of the sight's designers told newsmen that the Norden bombsight was indeed a factor in maintaining American air superiority over Europe and opined that the enemy would not be able to build anything equivalent to it during the rest of the war. He did not tell newsmen that the fabled bombsight had long since been found wanting in battle and had been technologically eclipsed by radar-guided systems.

On December 16, 1944, at Hitler's initiative the Germans began a counteroffensive in the Ardennes, with the goal of breaking through the Allied lines, decimating the mostly American troops, and retaking the key port of Antwerp. In support of that German counteroffensive, on that day the beginning salvos of what would eventually amount to 4,900 V-1 buzz bombs and almost 1,000 V-2 missiles were launched at the old port city. And the V-1s did not come in toward Antwerp at the 2,300-foot height that had been used when they had been sent against London but at 1,200 to 1,500 feet, which made the V-1s harder to detect. As important, the low approach caused the responding proximity-fuze shells' self-destruct mechanism to blast the shells apart near the ground, frequently injuring farmers and villagers.

Churchill, alarmed that the main way to counter the V-1s was not working properly, sent a fleet of long-range Lancaster bombers to wait in Cincinnati while the proximity fuzes' self-destruct switches were changed. "Accomplishing this [change] was easier said than done," Ralph B. Baldwin remembered about this moment. The self-destruct switch was activated automatically as the shell's rate of spin declined—a rate dependent on air pressure, altitude, muzzle velocity of the gun, the amount of wear and tear on the gun, and the firing angle of elevation. Using mechanical calculators, in two days Baldwin and colleagues completed the calculations for resetting the switches, different ones for each gun. Then the shells were remanufactured, loaded into the Lancasters, and flown across the Atlantic to Antwerp. Allied gunners were given a few days' practice firing the refigured proximity-fuze antiaircraft shells at wire-covered balloons, and then they had to get ready for the V-1s.

The German counteroffensive in the Ardennes area, some 250,000 men strong in three armies spread out over sixty miles, had penetrated through American and British lines. On the battlefields the Germans used nearly every available tank and all the new weapons that could be mustered, including the first squadron of sixteen jet fighter planes.

The weather over Antwerp was uniformly terrible. But radar would pick up an incoming buzz bomb twenty miles or three to four minutes from the port, allowing anti-aircraft firing to commence when a buzz bomb was one minute or about 12,000 yards away—still completely out of visual sight. "Often the buzz bomb was destroyed without having been seen by the crew," Baldwin recalled, while at other times, "it was seen only briefly as it bored low through scudding clouds." The results with the reconfigured proximity-fuze shells were spectacular. Of the 4,900 V-1s launched, about half went astray on their own, but of the 2,400 that would have hit the port area, antiaircraft defenses using the proximity-fuze shells destroyed all but 200. Antwerp remained intact as a port, enabling supply and resupply of the six Allied armies in the field.

To push back the "bulge" created by the German armies coming through the Ardennes, Allied ground forces under the command of General Montgomery used mostly their combined weight and matériel. They also had some new weaponry, such as new high-velocity ammunition made with a tungsten-carbide core, for piercing the armor of German Panzers; very little of that ammunition was available—just a few rounds per gun. But after the Joint Chiefs' authorization of the use of the proximity fuze over land, these were first used for that purpose. Bursting into shrapnel among massed German troops, proximity-fuze shells caused multiple deaths and injuries, and by their awfulness they also provoked panics, retreats, and surrenders. Tank commander General Patton reported to the head of Army ordnance that the "funny fuze" was a "devastating weapon. . . . When all armies get this shell we will have to devise some new method of warfare. I am glad you all thought of it first." Proximity-fuze shells—in the air against the V-1s aimed at the supply port of Antwerp, and on the ground against tanks and the massed German troops—played a key role in repelling the German counteroffensive of the Battle of the Bulge, and thereby kept the Allied juggernaut on track to end the war in Europe.

No further new Allied weaponry found its way to the battlefields against Germany, but new science-based weapons continued to emerge from the pipeline, in time for use in the last phase of the fight against Japan. Since 1942, the NDRC had been developing guided missiles in three different divisions: ordnance, instruments, and radar. Some were automatic once launched; others could be kept under the control of an operator while in flight. There was a television-guided glide bomb called "Robin"; it worked well but wasn't very accurate. It was replaced by several radar-guided bombs, each using a different source of information for the homing action: the Bat carried its own transmitter, the Pelican used a transmitter from the drop plane, the Moth homed in on the enemy's transmitter. The Roc used photoelectric cells to locate a target and then drop on it like a stone. There were so many new remote-controlled bomb devices that the ultimate one received the moniker of "Jag," for Just Another Gadget.

The television-guided Azon was the first to see battle, after more than $2 million in research by the labs of Remington-Rand, RCA, Fairchild Camera, Gulf Oil, Union Switch & Signal, a half dozen universities, and as many Army and Navy units. It took nearly five years and more than $2 million to develop because the problems were severe: when the difficulties of television image were resolved, the guidance and roll of the missile became the issue. Some tips were picked up from a German glider bomb sent to the United States by Colonel Boris Pash's team. In one test of the Azon, an engineer was horrified to see his own image on the target monitor, indicating that the missile was about to blow him up; it barely missed. By the spring of 1944 Azons were being used by planes based in the Mediterranean, blowing out locks on the Danube and the Avisio Viaduct near the Brenner Pass. In the invasion of Normandy, Azons made direct hits on eight bridges. In Burma they were twenty times as effective as conventional bombs, destroying narrow, single-track railroad bridges—twenty-seven in one day—and effectively shutting down Japanese resupply of their troops in that theater.

From the panoply of "birds"—self-propelled flying bombs—the Pelican emerged as the most promising weapon, developed jointly by the Rad Lab at MIT, the National Bureau of Standards, and the Navy.

But in June 1944 all six Pelicans dropped from a plane on a firing range missed the target; the Navy bucked the project back to development and switched to the Bat and to a homing system designed by Bell Labs that allowed the Bat to operate independently of the drop plane. In the spring of 1945 a squadron of Bat-equipped Privateer planes were sent to the Pacific; the missiles slung under their wings were immediately noticed by Japanese observers because they looked like miniature planes. Bats were launched from Privateers flying miles away from the targets—beyond the range of antiaircraft fire—at Japanese ships in or near harbors, with spectacular results: even when the targets maneuvered to get away, the Bats followed, one blowing up an ammunition ship, another severing the bow from a destroyer.

Perhaps the most peculiar remote-controlled-bomb project involved animals and the cereal company General Mills. An outsider suggested to the General Mills research director that dogs could steer submarine torpedoes. He called animal psychologist B. F. Skinner, then at the University of Minnesota, who thought a mechanical device could do whatever a dog might—but then decided that a brain would be a good thing in a guided missile and wondered if the animal he used regularly in his own experiments, the pigeon, could ride in a missile and guide it to a specific target such as an area of a city. The NDRC initially wanted nothing to do with the wild scheme, so General Mills funded some experiments itself, took movies of the results, and asked again for NDRC funding, which was then granted in the grand amount of $25,000. In the scheme the target image was projected on a ground-glass plate for the pigeon to peck at; if the pigeon pecked in the exact center, the bomb was on course, and if the pecking was off center, that meant the bomb was off course, and gyroscope controls would bring back the image to center line and the bomb to its proper course. As an official history relates, "The psychological factors had been handled with great ingenuity, and there seemed no doubt that the pigeons could be trained to do their job in spite of distracting noises and discomfort," but the project was abandoned anyway.

In early February 1945 one of the prime targets for American forces was going to be the island of Iwo Jima, a strategic outpost from which planes could fly most easily to bomb Japan itself. Iwo Jima was known to be heavily defended by Japanese troops, and, as an intelligence report put it, "stronger than Gibraltar." British and American chemical-warfare experts advanced a plan for subduing the island with mustard gas, while simultaneously jamming the Iwo Jima radio transmitter so that the garrison could not tell Tokyo of the nature of the attack. There was to be a message received from Washington by Admiral Nimitz, commander of the Pacific fleet, attributing the Japanese radio failure and the dead in the Iwo Jima garrison to an American "death ray." The plan even featured the removal of yellow identification bands from the gas shells so that gunners on the American ships would not know what was in the projectiles they would be firing. Nimitz was finally convinced that the plan was a good one, but he sent it to President Roosevelt for approval. It was rejected, with this note: "All prior endorsements denied—Franklin D. Roosevelt, Commander-in-Chief." Unlike Churchill, Roosevelt was unwilling to consider the use of chemical warfare, even when it was recommended to him by his most astute military chiefs and even when it would likely have saved American lives.

As the probability of defeat for the Third Reich grew, ever more desperate measures were conceived by the Nazis. Some were gigantic, like the thousand-ton tank that would be eight times larger than the biggest existing tank, the supergun known as the *England-Geschütze*, the six-engined tailless bomber Horten H-XVIII, and the superfast Daimler-Benz *Schnellstbomber*—none of which ever made it into production. Highly advanced, turbine-powered U-boats fueled by hydrogen peroxide, faster underwater than almost all Allied surface vessels, were produced in limited numbers starting in mid-1944 but Allied control of the seas prevented them from being used to advantage. After the Allied invasion of northern France, the U-boat bases had been transferred to Norway, and although there were more U-boats than ever before—463 by early 1945—Allied naval and air blockades and mining of the Baltic Sea prevented them from hindering the Allies' war plans.

At the other end of the size scale of weaponry in Germany were the very small planes, produced late in the war and designed essentially as suicide vehicles. In the fall of 1944 Hitler had decreed the creation of the *Volkssturm*, the people's storm, which would mobilize boys as young as sixteen. Around the same time he ordered the development of jet-powered and rocket-powered planes that could be constructed out of surrogate materials such as plywood. The "manned disposable equipment" would be sent up to do battle with Allied bomber squadrons, using machine guns until they were out of bullets, then ramming the Allied planes to bring them down. In some designs the boy pilot was to lie prone in the cockpit, so he could endure more gravitational force without blacking out. "If possible," a memo said, "they [the youth pilots] should proceed directly from glider training to mission He 162 flights without piston engine aircraft," in part to save precious fuel. Prototypes crashed and production was delayed because of Allied attacks on factories, but the ranks of the young would-be suicide pilots continued to grow, and the use of the "manned gliding bomb" was advocated—an adapted version of the V-1. Also proposed was a manned version of the V-2, with cockpit and wings on the big rocket, which von Braun boasted would be able to cross the Atlantic. At last, when fuel supplies in Germany were almost fully depleted, the plans were shifted to combat gliders, both unmanned and manned. "Thank God, this 'people's storm in the air' did not come into being," Adolf Galland later wrote.

But in early 1945 all the extant German aircraft companies, as well as individual designers, eagerly vied for contracts to design and sell the planes and flying bombs to the government, and worked overtime to fulfill their contracts, at a time when the general populace in the Third Reich was already convinced that the end of the Nazi regime was fast approaching. The designers' willingness to participate in desperate schemes reflected a belief that somehow science and technology would produce victory even at the last moment. The belief was encapsulated in a question that Albert Speer was frequently asked by various high-ranking military and civilian officials: when did the Führer intend to "apply the decisive secret weapon"? Often, when Speer said there was no such weapon, the officials refused to accept his word as final. Belief in technology had outrun all reason.

After the three-day Battle of Leyte Gulf in October 1944, the largest naval battle in history, the Japanese navy was finished as a significant military force. With its eclipse went Japan's ability to deliver food, oil, and raw materials through an increasingly effective Allied naval blockade. Japanese civilians were already starving, and the rations of the men of its armed forces were also curtailed to the point of compromising their physical strength. The Japanese air force had also been decimated and its remaining planes grounded most of the time for lack of fuel. In desperation, Japanese scientists devised ways of distilling aviation fuel from potatoes. But the kamikazes did not have to carry fuel for returning to base. On January 7, 1945, as American troops were attempting to retake the main Philippine island of Luzon, kamikazes put out of action two battleships, three minesweepers, and a handful of cruisers and destroyers.

Partly in answer to the kamikazes and to the atrocities routinely committed by the Japanese against Allied prisoners of war, and in an attempt to bring the war more swiftly to a close, massive American bombing raids on the Japanese home islands began in March 1945. Three hundred bombers dropped 2,000 "jellied gasoline" incendiaries on Tokyo on March 9, 1945, specifically targeting the densely populated area of Shitamachi, with its lath-and-clapboard buildings and small "shadow factories" that performed subcontract work for Japanese armaments firms. The resultant firestorm was larger than the one that had desolated Dresden, consuming sixteen square miles and killing more than 100,000 people. In the following week similar massive B-29 raids devastated parts of Nagoya, Osaka, and Kobe. Although such terror raids did significantly reduce the Japanese capacity to wage war, they seemed to have little effect on the government's resolve to continue that war, and they enraged the surviving populace, firming its will to fight on.

At the Rhine in March 1945, a team of American Chemical Warfare Service experts led by Lieutenant-Colonel Paul Tarr followed so closely on the heels of the Twelfth Army that complaints about their "nuisance" were sent to Washington. Then Tarr's team found an I.G. Farben factory at Ludwigshafen that was heavily damaged but still suspicious, and within hours determined that the plant had been manufacturing nerve gases. They also found the gases' inventor, Gerhard Schrader, convinced him to give them the formulas for sarin and tabun, and confirmed that supplies of tabun had been removed from stockpiles in Poland and loaded onto barges on the Elbe and Danube Rivers, for possible use in a last-ditch suicide attack to be directed from Hitler's retreat at Obersalzberg. They were also chagrined to learn from Schrader that Soviet troops had already overrun the Dyhernfurth plant manufacturing the nerve gases. Schrader did not know whether the Soviets had found the formulas and other secret manufacturing information, which had been hidden in a nearby mineshaft.[2]

Near Braunschweig in early April 1945, American infantry discovered and partially destroyed the remains of underground laboratories and wind tunnels before they realized that they had happened upon the Luftfahrtforschunganstalt Hermann Göring; later evaluation determined that this was probably the largest and best-equipped aeronautical testing laboratory in the world. Adolf Buseman's wind tunnels had been expanded so they could test supersonic craft, simulate altitudes of up to 50,000 feet, and take a thousand photos per second of engine workings. Buseman, most recently the inventor of the fixed-wing jet aircraft, was captured in its vicinity along with several colleagues. They seemed willing, even eager, to offer their expertise to the conquering Americans. Later, similar discoveries would be made at the Walterwerke, near Kiel, the German navy's research institute. "For the first two weeks," a member of the British discovery team later wrote,

> we found new weapons at the rate of two a day. Combustion chambers were hauled up from flooded bomb craters, key torpedo data dug up from underground, a miniature twenty-five-knot U-boat salvaged from the bottom of a lake . . . and there were prototypes of dozens of new and ingenious weapons: long-range guns, minesweeping devices and jet-powered grenades.

A Type XVII submarine was rebuilt and shipped to the United States, and a Type XVIII was similarly rebuilt and sent to Great Britain. But because the Soviets had taken the V-2 rockets from Blizna and had reneged on an agreement to share the booty from a

[2]The Soviet forces had found the formulas and had taken them back to Soviet laboratories for analysis and reproduction.

German torpedo-research center at Gdynia, the British and Americans at Kiel were instructed to tell the Soviets that all the important objects at Kiel had been sunk or destroyed by the Nazis.

American forces captured Heidelberg, whose university was an important outpost of German science; there Goudsmit had to confront for the first time a physicist whom he knew personally, Walter Bothe, not only an old colleague "but certainly my superior as a physicist." Bothe greeted him warmly, showed him all his papers published during the war and his cyclotron, but balked at divulging what he thought was war-related information. Gerlach, another friend of Goudsmit's, younger than Bothe, looked "haggard" but was happy to see Goudsmit and proud that he had "done his best to save what he could" of German physics from the depredations of the Nazis. Then there was Lenard, who because of his Nazi connections held the top physics post at Heidelberg; when Lenard was captured by ordinary soldiers, he informed them that he was Germany's greatest physicist. The soldiers asked Goudsmit what to do with Lenard, and he told them to "ignore him. . . . This, for a Nazi, was a greater punishment than being tried at Nuremberg."

But Heisenberg, von Weizsäcker, and whatever remained of the Nazi nuclear program were still at large, deep in Germany.

On April 12, 1945, President Franklin D. Roosevelt died of a cerebral hemorrhage at Warm Springs, Georgia. Though Americans had known that Roosevelt was ill, they, the Britons, the populations of the other democracies, and many people in the countries subjugated by Germany and Japan felt keenly the loss of the American leader. But in Berlin, Hitler was overjoyed at Roosevelt's death, telling Speer, "Here we have the miracle I always predicted. Who was right? The war isn't lost." The Japanese showered leaflets over American troops trying to seize control of Okinawa: "The dreadful loss that led your late leader to death will make orphans on this island. The Japanese

Special Assault Corps will sink your vessels to the last destroyer. You will witness it realized in the near future." Soon Allied ships were attacked by waves of kamikaze and *kikusui*, "floating chrysanthemums" of frogmen tied to suicide bombs.

In the battle for the island of Okinawa, the *Yamato* itself became a suicide ship, carrying only enough fuel to reach the island, not to return to its base. Attacked by Allied torpedo bombers before it could come to the aid of the island's defenders, the *Yamato* was sunk along with more than 3,000 of her crew.

The Japanese had developed a version of the proximity fuze and used it to devastating effect against an American squadron of B-29 bombers stationed at an airfield in Saipan, destroying most of the bombers on the ground and delaying a new wave of terror bombings scheduled against Japanese cities. Some 20,000 proximity-fuze shells were manufactured in Japan and stored in the home islands, for possible use to repel an Allied invasion. Were these proximity fuze shells to have been used by Japan against Allied ground forces—in the same manner that the Allies used them against German forces in the Battle of the Bulge—they would likely have produced casualties numbering in the tens of thousands. Less serious as threats but also awaiting an Allied invasion were hundreds of *baka*, rocket-powered suicide gliders, and a prototype of a Japanese death ray. After years of fictional fantasies, an actual death-dealing device had been made. It was a direct progenitor of today's microwave oven. But while today's oven cooks—kills—by bouncing the waves around inside an enclosed space, the Japanese death ray aimed its microwave power outward, with reflectors focusing it on a target; the prototype had been successful in killing a rabbit and other small animals at ranges of up to a hundred yards.

The Germans also contemplated a death ray. Shortly after Roosevelt's death, and as Allied and Soviet forces were making rapid progress toward the center of the Nazi empire, Speer was braced by Dr. Robert Ley, who told him that "death rays have been invented! A simple apparatus that we can produce in large quantities. . . . This will be the decisive weapon!"[3] Speer agreed to have Ley take over the project and assume the title of Commissioner for Death Rays. Ley ignored the sarcasm and took the job, but later claimed he was unable to complete the project because he lacked certain tubes and circuit breakers.

The German "death ray" would not have turned the tide of the war in Europe; perhaps nothing could have done so. But a glimpse of how another technologically advanced weapon might have altered the course of the war was provided in mid-April by a demonstration of the power of jet planes. As a phalanx of B-17 bombers flew toward Berlin, they were attacked by six Me 262 jets maneuvering at speeds up to 700 miles per hour and using the deadly R4/M missiles; fourteen of the Allied bombers were destroyed, and the raid was blunted. Had several hundred such jets been available to Germany at the time of the Normandy invasion—and they could have been produced—Allied air superiority would have been compromised, and the fate of the invasion might well have been different.

In late April 1945, as German defenses were falling apart, Göring publicly announced his willingness to immediately succeed Hitler as chancellor. From his bunker, Hitler replaced Göring in the line of succession with Admiral Dönitz, the former submarine chief who had risen to become head of the German Navy. The change in successor was one of Hitler's last decrees. On April 30, 1945, as Soviet troops were entering Berlin and as Allied forces were liberating Dachau and Buchenwald, Hitler committed suicide.

Leadership of the government passed only partially to Dönitz, while Goebbels became chancellor. By May 2, troops of the Third Reich were surrendering in wholesale lots to various Allied commanders. In southern Germany, Wernher von Braun and Walter Dornberger identified themselves as rocket experts when they gave

[3]The designs for Ley's "death ray" probably had nothing to do with the Japanese version.

themselves up to American forces. As they had hoped, they were soon debriefed, taken out of the country, and started on their way to continuing their work in the United States. Heisenberg, Hahn, von Weizsäcker, and other nuclear scientists came out of redoubts and retreats and gave themselves up. Surrenders by the various German armies and individuals continued until the final surrender of all forces, authorized by Dönitz and signed by General Alfred Jodl at 1:41 A.M. on the morning of May 7, 1945.

After the defeat of Germany the great decision facing the Allies was whether to mount a land invasion of the Japanese home islands. On this issue Vannevar Bush and Admiral Ernest King—who so frequently had held opposing views—agreed: because the Allies already enjoyed naval and air superiority, they felt that an invasion was not necessary to bring Japan to the peace table. While some planners advocated that the Allies simply wait and starve the Japanese home islands into a surrender, others wanted to use the power of technology to force that conclusion. They argued that the atomic bomb would more quickly bring the war to an end and foreclose any possibility of an Allied invasion. Had there been no bomb, the pros and cons of slow starvation might have been more fully considered. But there *was* a bomb, even though, in the late spring of 1945, there were still many high-ranking military and civilian officials who did not believe that this bomb could be successfully exploded. When President Truman was first briefed about the bomb by Bush and Admiral William Leahy, while Bush told the president that the bomb would work, Leahy said of the Manhattan Project, "This is the biggest fool thing we have ever done. The bomb will never go off, and I speak as an expert in explosives."

Two months after that briefing, at 5:43 A.M. on July 16, 1945, the first atomic bomb in the history of the world was exploded at Alamogordo, New Mexico. Ten miles from ground zero, the principal scientists—along with Bush, Conant, and Groves—lay in modest shelter, shielding their eyes from the blast with pieces of dark glass. After the explosion and shock wave, Bush and Conant shook hands. Just then President Truman was attending his first conference with Churchill and Stalin, at Potsdam, a suburb of Berlin that had been heavily bombed by the Allies. General Groves conveyed news of the atomic-bomb test to Secretary of War Stimson, at the Potsdam conference, in a transparent code: "Operated on this morning. Diagnosis not yet complete, but results seem satisfactory and already exceed expectations." Truman was exhilarated and relieved by the news, and Churchill called it "a miracle of deliverance." Neither leader fully understood what an atomic bomb might do to Japan—or to the larger world. Keeping the news from Stalin, the two Western leaders discussed the weapon in terms of being able to avoid, in the proposed invasion of Japan's home islands, the terrible blood costs that had been paid in the taking of Okinawa. Churchill expressed hope that the Japanese "might find in the apparition of this almost supernatural weapon an excuse which would save their honour and release them from their obligation of being killed to the last fighting man."

In the midst of the Potsdam conference, Churchill was called home because of the general election, which threw his Conservative Party out of office. The Labour Party's Clement Attlee became prime minister, and he returned to the parley at Potsdam. Truman waited until Attlee's arrival, and on July 24 informed Stalin that "we have perfected a very powerful explosive which we are going to use against the Japanese and we think it will end the war." Truman did not use the term "atomic bomb." Stalin nodded at the news but did not tell Truman that the Soviets already knew quite a bit about what had gone on at Los Alamos. Later, in private, Stalin sent a message to Moscow that the Soviet nuclear scientists must "hurry things up." On July 26, 1945, Truman, Stalin, and Attlee issued a declaration at Potsdam calling on Japan to accept "unconditional surrender" or face "the inevitable and complete destruction of the Japanese forces, and just as inevitably the utter devastation of the Japanese homeland."

There is no evidence that any Japanese scientist who learned about this warning—which itself was not widely disseminated in Japan—conveyed to the government that the "utter devastation" promised by the Allies might come from an atomic bomb.

The decision to use the atomic bomb against Japan was neither easily reached nor unanimous. Vannevar Bush was among those who argued for it, not because it would prevent Allied soldiers from dying needlessly in an invasion of Japan but because he thought it would result in fewer Japanese deaths than would starving the populace into surrender or using incendiary bombs against more cities. In the end what appears to have tipped the balance toward the descision to drop an atomic bomb on Japan was the supposed deterrent effect it would have on Stalin: scaring the Soviet Union from further territorial aggrandizement in Europe and halting its potential advances into Manchuria and other territories then controlled by Japan.

Ten days after the Potsdam warning to Japan, the first atomic bomb, known as "Little Boy," was loaded onto the *Enola Gay,* which headed through the early-morning hours toward Hiroshima. The automatic controls for dropping the atomic bomb took over at 8:15:02 A.M. on August 6, 1945, and at 8:15:17 opened the bomb bay doors on the *Enola Gay.* The bomb began to drop through the air from 30,000 feet. At 1,800 feet the barometric-pressure device triggered the detonator. The fireball of the blast was hotter than the surface of the sun, burning everything in its path, melting granite, imprinting the shadows of obliterated people and objects on the few walls that survived. As a result of the shock wave and blast, ten square miles became a wasteland and 130,000 people died, most of them instantly, the rest within days.

When the news was received at Los Alamos, the Manhattan Project team members applauded and clapped one another on the back. President Truman, aboard a ship in the Atlantic, exclaimed, "This is the greatest thing in history." A White House communiqué the next day announced the dropping of "an atomic bomb . . . harnessing the fundamental power of the universe," estimated at equivalent to 20,000 tons of TNT. "If they [the Japanese] do not now accept our terms, they may expect a rain of ruin from the air, the like of which has never been seen on this earth."

The Japanese public was not informed by its government of the nature or extent of the damage at Hiroshima, and the government did not initiate surrender overtures. Shortly its ambassador to the Kremlin was called in by Molotov and read a declaration of war by the USSR. That night the B-29 *Bock's Car* took off with the "Fat Man" plutonium bomb in its bay, aiming for Kokura and its large army arsenal. At dawn on August 9 Kokura was enveloped in thick clouds; the B-29 was diverted to Nagasaki. Simultaneously Soviet troops overwhelmed Japanese defenders at many positions on the Manchurian border. Shortly before 11:00 A.M. on August 8, 1945, the second American atomic bomb was dropped on Nagasaki; the surrounding hills absorbed some of the blast and limited the damage, so that the total number of those who died as a result of this bomb was 45,000 people.

That evening debate raged in the Japanese cabinet as it met in the emperor's bomb shelter. The ministers were evenly divided between those who argued for accepting the harsh terms of the Potsdam declaration, and those who were for continuing to fight—as the war minister put it, to "find life in death," presumably as an entire nation of kamikazes. Emperor Hirohito made the decision: the terms of Potsdam were "unbearable," yet to avoid the "prolongation of bloodshed and cruelty . . . the time has come when we must bear the unbearable."

Next morning the decision to accept the Potsdam terms was made, with the proviso that the emperor must maintain his position as sovereign ruler. American and British leaders soon agreed to accept these terms; Stalin did not, so that his armies could advance farther into Manchuria. While the issue was unsettled, American military officials ordered the readying of the third (and last available) atomic bomb for dropping on Japan.

It was not used. By August 15, 1945, all belligerents had accepted the peace terms, an incipient revolt in Tokyo against the surrender had failed, and a tape recording by the emperor was played on Japanese radio. The Allies' use of the "new and cruel bomb, the power of which is indeed incalculable," the emperor said, made it necessary to accept the Potsdam demands, since the continuation of the fight "would not only result in an ultimate collapse and obliteration of the Japanese nation, but would lead to the extinction of human civilization."

World War II was over.

Epilogue: Science, Technology, and the Postwar World

Today's headlines about international terrorists coordinating their attacks by computer and wireless telephony, the need for "Star Wars" antimissile defense systems, battlefields controlled from satellites in space, and fears of guerrilla warfare using biological and chemical agents of sabotage are all a reminder that science and technology have taken over military considerations to a degree almost unimaginable at the close of World War II. In today's world, all war is science-based, whether waged by technologically sophisticated countries, or by terrorists able to turn a highly developed country's dependence on its technology against its own institutions.

These conditions have come about in large measure because of the intermingling of science, technology, and military endeavor that began in earnest during World War II. During that war, a lot of scientific work was done at each government's behest—some of it good science, some shoddy, but most of it simply utilitarian and technologically oriented. The Allies, principally the United States and Great Britain, mobilized and used their science-based resources well, and they benefited from doing so in many ways that contributed to their ultimate victory. As for the Axis countries, there is general agreement that their science-based resources were not used well and that this misuse was a factor in their defeat.

A myth prevailed for a long time: that Germany had lost the scientific war because it had expelled its Jewish scientists in the early 1930s, after which the intellectual capacity of Germany's science was fatally diminished. Certainly the work of the many refugee scientists in various war-related efforts in the West, notably the production of the atomic bomb, was evidence supporting that view. But today, after additional years of retrospect, it has become clear that Germany's war fortunes were more adversely affected by Hitler's decade-long refusal to see Germany's scientists as a national resource or to properly mobilize them. After Germany's expulsion of the Jews, a strong contingent of first-class scientific minds remained in the Third Reich, but they were prevented from contributing to Germany's military prowess early enough in the war and in ways that could have made a difference in the military outcome. German scientists and technologists had ideas that could have advanced radar, submarines, jet planes, antiaircraft rockets, guided missiles, and other new weaponry to the point of tipping the balance in the war, but their expertise was never fully engaged. Moreover, the scientific effort was hampered by the Nazi hierarchy's magical thinking, expressed in the preference for developing the huge but ineffective V-2 rockets while neglecting the Waterfall antiaircraft rockets and jet planes that could have successfully countered the Allied bombing raids.

The science-based weaponry that Japan possessed at the war's outset—some of it quite advanced, like the long-lance torpedo—was not much improved by war's end. The major factor holding back development was the arrogance of the military, expressed in its ignorance of how to mobilize scientists or to critique what scientists were or were not doing to further the war effort. After the war, recognition of this wartime failure spurred the new government to rely heavily on science and technology to economically rebuild Japan and to achieve by economic hegemony what Japan had never accomplished militarily: the dominant place in the Asian world.

Tomorrow's electronic battlefields began with the avidity of the victor governments' attempts, at the close of the war, to capture and make use of the talents of German scientists and technologists. The most famous and the largest corps of new immigrants to the United States was the Peenemünde rocketeers led by von Braun, Dornberger, and Arthur Rudolph, a group that eventually included more than four hundred men and their families: they became the nucleus of the American guided-missile program and later of the space program.

Wernher von Braun later worked as a design consultant for and became identified with the Walt Disney Company and its squeaky-clean image; and in many other ways the German rocketeers transplanted to the United States were regularly portrayed to the American public as having been only tangentially involved with the excesses of the Nazi regime. They were far from blameless on that score; as with the research groups involved with nerve gases, or with jet-plane design and manufacture, or with advanced submarine design, the rocket-scientist corps contained some obvious and long-term Nazis like Rudolph and plenty of men whose actions between 1933 and 1945 would have, by American Army rules, rendered them unfit to be granted permission to become citizens of the United States or—in some instances—would have earned them indictments for war crimes. American governmental agencies, principally the military and the intelligence services, knowingly falsified these men's records and deterred inquiries into their backgrounds, believing it necessary to gloss over the Nazi affiliations and past transgressions of those Germans whose scientific and technological expertise were considered essential to the future security of the United States, men who were also being courted or coerced to work for the USSR.

Other instances in which American authorities overlooked Nazi pasts were less deliberate and resulted from an overreliance on wrong recommendations made by former German scientific administrator Walter Osenberg. Sitting in Paris with his wartime files of 15,000 Germans involved in scientific projects—originally compiled for Göring, with the aid of the Gestapo—Osenberg, at the behest of the American military, culled and annotated a list of those men he labeled as the best German scientists, those who merited protection by the United States. Unbeknownst to his Paris supervisors, Osenberg gave good marks to many Nazis and sympathizers while awarding bad marks to scientists with whom he had feuded or who had not avidly cooperated with the regime—for instance, assessing Nobelist Otto Hahn and nuclear researcher von Weizsäcker as of "negligible value." When Vannevar Bush saw Osenberg's list, he pointed out that Hahn and von Weizsäcker were "intellectual giants" and castigated the military for ignorance of "elementary information on Germans whose names are as well known in scientific circles as Churchill, Roosevelt, and Stalin are in political circles." Further bending of the lists may have been done by Donald Maclean, the British intelligence agent who was a mole for the Soviet Union.

Such mistakes might well have been expected in the frantic recruiting environment that pitted American efforts against similar ones mounted by Great Britain, which offered scientists a variety of new venues in the Commonwealth in which to resettle; France, which offered better pay; and the Soviet Union, which offered all scientists contracts even if their names were on Allied automatic-arrest lists.

This recruiting by the Western and Communist powers was among the opening actions of the Cold War. Within five years after the end of World War II, rival groups of German technologists working for the United States and the USSR would produce jet fighter planes and other weapons systems, based on designs drawn during the waning days of the Third Reich, that would see combat against one another on the Korean peninsula.

One group of German scientists whom the United States government specifically did not recruit (though Bush and others recommended it) were those engaged in nuclear research: General Groves did not want any of them near the Manhattan Project, which, in addition to fashioning the first atomic bombs, was in 1945 at work on the "super," the hydrogen bomb.

Ten of the most important German nuclear scientists, including Gerlach, Hahn, Harteck, Heisenberg, von Laue, and von Weizsäcker, were interned at war's end in Great Britain, at an isolated country estate near Cambridge called Farm Hall. Their conversations were surreptitiously recorded, and the tapes remained classified for almost fifty years.

The Farm Hall detainees, thinking they were not being overheard, posed for themselves frank questions about their pasts and futures. Only two of the ten had been members of the Nazi Party, but only three had avoided joining some Nazi-affiliated organization; the group considered drafting a statement that they had been "anti-Nazi" during the war and in other ways sought to emphasize their distance from the depredations of the Nazis. In fact, they all had cooperated with the regime, though to varying degrees, rather than opting out of its embrace or actively resisting it. Unlike the rocket scientists, who had knowingly used concentration-camp labor to build missiles, the nuclear scientists for the most part had no direct touch with the more horrific excesses of the Nazis—but they had not refused the perquisites awarded to those who actively cooperated. At

Farm Hall, to ensure their future positions, they concocted myths to disguise their past intimacy with the Nazis: they contended they had been apolitical, had worked on the bomb project only to keep German physics intact, and had really been attempting to achieve international cooperation to prevent all countries from using nuclear weapons.

On August 6, 1945, when the radio carried to Farm Hall the announcement that the atomic bomb had been dropped on Hiroshima, Heisenberg and others did not at first believe the report. Then the discussion turned technical—whether or not plutonium had been used and how many kilograms of fissionable material had been necessary—and after being pressed by Hahn for answers, Heisenberg did some calculations and gave a lecture showing what he now understood of how the bomb worked and how it had been produced. The impromptu lecture demonstrated both the errors the German nuclear scientists had made and how quickly they could have come to the proper understandings had they earlier corrected their mistaken assumptions. It also made clear that the limited nature of the Nazi atomic-bomb program had been due more to the unwillingness of the upper hierarchy to commit the large amounts of money and resources needed to make a bomb than to anything that the scientists had or had not done.

Many earlier émigré scientists believed that their former colleagues who had remained in Germany bore some blame for the horrific deeds of the Third Reich. Their attitude was well expressed in a letter from Lise Meitner in Sweden to her longtime collaborator, Otto Hahn, on June 27, 1945.[1] In her letter Meitner charged that the "misfortune of Germany" was that "all of you lost the measure of justice and fairness," that Hahn, von Laue, Heisenberg, and the others with whom she had worked (and corresponded during the war) had

[1]The letter was sent to him in Berlin, but he was not there, and Meitner later wrote that it "never reached him."

known of the "terrible things . . . being done to the Jews" and had not spoken out against them. Even worse, in her view,

> All of you also worked for Nazi Germany and never tried to engage even in passive resistance. To be sure, to soothe your conscience you helped someone in trouble now and then, but you let millions of innocent humans be murdered and no protest was raised. . . . You never did have sleepless nights. You didn't want to see it, it was too inconvenient. . . . You bear responsibility for what happened because of your passivity.

Of the 15,000 scientists and engineers involved in war-related work in Germany, a thousand or so were relocated to the Allied countries to continue their work, and it is estimated that a like number went voluntarily or involuntarily to the Soviet Union for the same purpose. The Allied authorities responsible for drawing up the lists of war criminals considered the remainder of the scientific workers—the practical physicists who developed guns, planes, submarines, mines, and other weapons; the chemists whose efforts to produce substitutes for fuel and rubber prolonged Germany's ability to make war; the biologists who increased Germany's food supply—as bearing no responsibility for the tasks they had been given to undertake, and therefore were not to be prosecuted. But they also applied the same exculpatory logic to other German scientists who might well have been held legally culpable for contributing to the large-scale crimes, such as the biologists, eugenicists, and psychologists who developed the criteria for selecting people to be sterilized or sent to the death camps, and the engineers and chemists who designed and built the mass-extermination machinery. The only group directly accused by the Allies of war crimes was a handful of physicians, whose depredations in connection with the concentration camps and experiments on inmates were too obvious to ignore. Speer and Göring, who had supervised scientific research, were tried at Nuremberg, though not on charges having to do with science; Speer admitted his guilty participation in the excesses of the regime and went to prison, but the field marshal remained unrepentant and committed suicide before full judgment could be passed or retribution exacted.

The Nuremberg war crimes trials were supposed to go hand in glove with a de-Nazification program. That did not work very well. In postwar West Germany some scientists who had cooperated with the Nazis were ostracized, while only a few of those who had suffered at the hands of the regime or who had opted out rather than cooperate were reinstated and elevated to senior positions at universities and institutes. West Germany made no attempt to expunge from the code books the 1933 Law for the Restoration of the Civil Service, or to reinstate scientists who had been dismissed under that law, or to compensate those who had lost their jobs by reason of that law.

In regard to Japanese scientists, the pattern was much the same: scientists who had been engaged in research that the United States coveted, such as people associated with biological and chemical warfare—who might well have stood trial for their experiments on living human subjects—were extensively debriefed and in a few instances allowed to emigrate to the United States, though most were permitted to return unhindered to civilian life in Japan. Although Allied tribunals brought 2,000 cases against 5,700 individuals in Japan, only two were for cases of medical experimentation, done on Allied prisoners. Chinese former prisoners of war who survived Ping Fan were outraged when the Allies did not indict even a single member of the Japanese military or medical establishments for the crimes committed there. The Soviet authorities did put on a show trial at which they convicted a dozen of Major Ishii's Ping Fan subordinates. Major Ishii himself was permitted by U.S. officials to go free in exchange for turning over his records and other information; to obtain immunity from prosecution, Ishii exploited the American fear that his expertise might go to the Soviets.

In Japan, to a greater degree than in postwar Germany, scientists who had done wartime research for the government attracted little opprobrium and during the American occupation rather easily found work in the universities, institutes, and companies. A report to the United Nations suggested why:

> At a relatively early period [of the occupation], it became clear that the democratization of Japan would depend in large measure upon economic rehabilitation of the country. It was clear . . . that science and technology would have very significant roles to play in any rational scheme. . . . Hence, at an early stage . . . emphasis shifted from the surveillance function to a policy of friendly guidance designed to increase the ability of Japanese scientists and technologists to contribute to the economic rehabilitation.

Later, in biographical sketches, many leading Japanese scientists simply left out of their histories the positions they had held during the war, some leaving blank an entire decade or more.

Allied postwar trials in Japan and Germany convicted a few scientists of crimes against humanity, while at the same time the governments of the United States and of other Allied countries eagerly sought and collected the data these scientists had amassed, characterizing it as important advances in understanding, no matter what the circumstances or methods involved in first obtaining it. Such contradictory actions by the Allies rendered even more faint the line between use and abuse of science in wartime.

Leading scientists who participated in Allied war-related research and development, particularly in Great Britain and the United States, reaped benefit from their cooperation. Although it has been argued that the Manhattan Project and the MIT Rad Lab were deliberately staffed with the brightest scientific minds, those people believed to be destined to make important discoveries, it is nonetheless striking how many Nobelists of the war and postwar years had their tickets punched for stardom at the Manhattan Project or the Rad Lab. Among them were I. I. Rabi (Nobel Prize, 1944), Percy Bridgman (1946), Felix Bloch and Edward Purcell (1952), Willis Lamb (1955), Owen Chamberlain and Emilio Segrè (1959), Richard Feynman and Julian Schwinger (1965),[2] Hans Bethe (1967), and Luis Alvarez (1968). American Nobelists Robert Hofstadter (1961) and William Fowler (1983) worked on the proximity fuze during the war, though their prizes were awarded for other work. Several Nobels to Americans may be traced directly to research done during the war: the development of the transistor by William Shockley, John Bardeen, and Walter Brattain (1956); and the Nobel to Lamb and Polykarp Kusch (1955), who built on their wartime experience with microwave radar and vacuum-tube construction to develop techniques for passing microwaves through hydrogen atoms to learn more about them.

Great Britain's postwar Nobelists also featured scientists who had worked actively with the government during the war, such as Ernest Chain (1945); Edward Appleton (1947), who oversaw many research labs; P. M. S. Blackett (1948); John Cockcroft (1951); Peter Medawar (1960); and Martin Ryle (1974), an astronomer who worked at TRE. Soviet scientists who put aside their prewar basic physics and chemistry studies and toiled assiduously during the war for the government also earned a large share of the postwar Nobels: Nikolai Semenov (1956); Cherenkov, Tamm, and Frank (1958); Landau (1962); and, belatedly, Kapitsa (1978). Only one Nobel to Soviet researchers could be linked directly to war-related work: the physics prize to Basov and Prokhorov (1964) for development of the maser, which emerged from their work on radio-wave propagation.

In the postwar period Germans were generally rewarded with

[2]Shared with Sin-Itro Tomonaga, who had worked in isolation in Japan during the war.

Nobels only for research done in the 1930s, such as Diels and Alder (1950), whose work formed the basis of much of the postwar industrial-chemicals industry. Their collaboration with the regime was excused on the grounds that because of their work's practical applications they seemed to have had little choice in the matter. More controversy surrounded the awarding of the 1973 Nobel Prize in physiology to Konrad Lorenz, who had gone out of his way to work with the Nazis, while his corecipients, Karl von Frisch and Niko Tinbergen, cofounders with Lorenz of the science of ethology, had suffered at the hands of the regime, Frisch because he was part Jewish, Tinbergen under the occupation of the Netherlands.

Cherwell's influence in Great Britain faded in the postwar period when his patron, Churchill, was out of office. That of men like Cockcroft, Reg Jones, Zuckerman, and Medawar, who had risen to responsible positions during the war, continued through several changes of government, affecting policy for the next quarter century. Tizard remained on the sidelines in academia. In the postwar era J. D. Bernal's 1939 book, *The Social Function of Science*—heavily influenced by Communist thought on the proper relationship of science to the state and the need to plan all research—became a bible for the reorganization of science in Belgium, the Netherlands, and other small European countries. Frédéric Joliot, in part because of his Resistance work in the war, emerged in the postwar era as the most important figure in charge of resurrecting science in France in all its configurations; Joliot's expertise in nuclear science encouraged the French to develop their own nuclear weapons and power plants.

All the European countries, Great Britain included, had to apply the largest portions of their national budgets to rebuilding, not to scientific research. The United States, whose resources were not similarly affected, in the postwar period consolidated the position it had attained during the war, as the world's most important and most productive center for science and technology. It did so despite the frustration of Vannevar Bush's initial attempts to establish the National Science Foundation. Truman vetoed a compromise bill on the NSF in 1946, and by 1950, when the NSF was finally chartered, its purview had been seriously diminished from what Bush had originally imagined, an entity that would have controlled military, medical, and basic research. Bush grumbled that the military had taken over too much, but actually the military had now adopted his wartime proposals for its own use, including a permanent partnership between the military and civilian scientists employed at nonmilitary establishments. For instance, Admiral Bowen, whom Bush had shunted aside in 1940 because he refused to condone military-related research being done by civilians, emerged in 1946 as the czar of the new Office of Naval Research, a major scientific-development institution within the armed services, employing a large number of its own scientists, and he also took the lead in urging the awarding of Navy contracts to civilian entities, which soon amounted to $70 million a year in research, making the Navy, as *Newsweek* put it, into "the Santa Claus of basic physical science."

Some scientists who wished after the war to return to the pure-science pursuits they had been engaged in before the conflict began could not do so. The most egregious case was that of Nikolai Timoféeff-Ressovsky, the Russian-born geneticist who had worked in Berlin throughout the Nazi period. He had remained mostly independent of the Nazi embrace but could not avoid having to put his expertise to government use in such matters as developing better gas masks. By late in the war he had become chief of a laboratory and a secret foe of the regime—his eldest son had engaged in underground anti-Nazi activities and had died in a concentration camp. As Russian troops approached Berlin, Timoféeff brushed off the pleas of friends from the West that he join them, arguing that he was now responsible for his institute's staff and would be better able than anyone else to intercede with the Russians on their behalf and keep the institute intact. He succeeded in doing those things, but only for a short time, after which the institute's equipment was sent to Moscow and Timoféeff was arrested and imprisoned in Moscow on the charge of having been a spy. In prison he organized a Scientific and Technical Society, as he did again when transferred to a camp in the gulag—a society that, according to Zhores Medvedev, saved him and his fellow scientists from intellectual death. His physical health deteriorated, and he nearly lost his eyesight before his expertise was deemed essential to the Soviet nuclear effort, part of which involved the study of the effects of radiation on living tissue. Though he still yearned to do basic genetics research, he became director of the Soviet laboratory in the Urals devoted to the radiation work. He clashed with the Lysenkoites, whose opposition continued to control the study of genetics even after the death of Stalin, preventing Timoféeff from receiving recognition for his work until the late 1960s.

At the other end of the scale was Merle Tuve. The principal developer of the proximity fuze during the war, after it he did not continue with military-related research—which he could easily have obtained—or return to either the exploration of the upper atmosphere that had occupied him in the 1920s or the nuclear research that he had helped pioneer in the 1930s. Rather, he turned his restless mind to a new field of endeavor: the earth's crust; using war-surplus mines and depth charges, he set off underground explosions and measured their paths, enabling him to map the earth's crust down to 125 miles—an accomplishment that mirrored his earlier radio-wave mapping of the atmosphere 30 miles above the earth.

W. A. Noyes, the chemist who was a professor at the University of Rochester, a division leader of the OSRD, and a member of the ALSOS team during World War II, spoke for many scientists when he wrote just after the war that the chemical research undertaken by American scientists in wartime "involved very few new ideas." The country's drawing on its "scientific resources to prosecute the war" contributed to the success of the Allies, he was proud to say, "but it must be kept in mind that the work was mainly of a developmental character and involved little fundamental research. Old ideas were extended to new problems, many new techniques were devised which will be useful in fundamental research. Yet, basically, science was not greatly advanced by all this effort." Noyes pleaded with the government and the public not to "feel that all work from now on should be of an applied character. For the good of the country and for the good of science we must get men back to the laboratory with leisure for thought and for fundamental experimentation."

Noyes's notion of scientists doing only basic research without regard to practical considerations was an ideal that had long since been overtaken by events. In the postwar period all of science in the United States—basic as well as applied, microbiology as well as physics—became increasingly dependent on government support, which markedly changed its character. Self-funding of research by private institutions was drastically reduced; each investigator's work had to earn outside funding in order to proceed, which meant that its goals and relevance to social benefit had to be made overt and stated in advance, a need that (as Pavlov had long ago predicted) narrowed the scope and reach of the research, often tipping the balance away from simply trying to expand a field's base of knowledge. That the applying of standards of practicality to basic research might be shortsighted was emphasized by Harvard's John van Vleck, the 1977 Nobel Prize–winner for physics, when he pointed out that "None of us who worked in molecular spectra in the 1920s dreamed that two decades later some of the results might have military significance, and four decades later important applications to radio astronomy and astrophysics."

In the postwar climate the high budgets for military research fed into the thirst for relevance by providing large pots of money for only those scientific projects that could be billed as advancing an eventual practical goal while in the pursuit of basic knowledge. Growing cadres of scientists and technologists toiled on projects for "defense contractors" and at universities such as Caltech and MIT. The Cambridge school became the largest nonindustrial defense contractor, with seventy-five separate contracts worth $117 million annually. Oak Ridge leader Alvin Weinberg quipped that it was "increasingly hard to tell whether MIT is a university with many government research laboratories appended to it or a cluster of government research labs with a very good education institute attached to it."

Added jobs for scientists and technologists came from the new commercial-electronics industry, which was based on the $2 billion invested in radar development by the Allies during the war, and from the new computer industry, which similarly emerged from the hundreds of millions of dollars spent on developing computers by the military, the intelligence services, and the nuclear-bomb program. The GI Bill, which paid for college for half of the 17 million men and women who served in the U.S. armed forces, helped to create many more technologists, graduates of already established colleges and of new, two-year colleges, whose training programs were aimed directly at readying people for employment in the electronics industries.

The total amount of scientific research skyrocketed, especially in the United States, which in the postwar era eclipsed Europe in terms of the sheer weight of scientific endeavor, and this contributed to the ascendance and acceptance of science and technology in American society and culture. It found expression in the plethora of electronic gadgets in the home and office, in the setting and achieving of the goal of placing a man on the moon within a single decade, and in the cornucopia of new drugs and medical treatments that steadily increased people's average life span and the quality of their health.

There was a military downside to the fact that science and technology took over Western society so completely that they became synonymous with advanced civilization. Their influence on Western military planning and armament was such that the armed services came to rely so heavily on sophisticated sensing and guidance technologies that they discounted other elements of combat. The mismatch of forces during the conflict in Vietnam, featuring the technologically-rich American armed services fighting the less-well-equipped North Vietnamese, and the eventual victory of the Communists—though attributed to political factors—began to expose the West's error in overdependence on scientific weaponry and its failure to properly assess the military power of ideology and guerrilla warfare. Decades of acts of terror by religious and nationalistic fanatics against Western targets, culminating in the 2001 destruction of the World Trade Center towers in New York City, have not yet persuaded Western military planners to balance their reliance on technology with more emphasis on developing non-mechanized methods for countering these new enemies.

During World War II, most of those in the democracies engaged in scientific and technological work for the military viewed the experience in a positive way, believing they were doing what was required to defeat the Axis and, not incidentally, to make the world safe for the practice of science. When the war was over, the GIs shed their uniforms and took up civilian clothes; the scientific wizards did not shed their robes but did return in like manner to civilian pursuits.

Some had second thoughts, doubts about having permitted their expertise to be harnessed to political and military goals. Chief among the doubters were some who helped develop the atomic bomb. The bomb's vast capacity for destruction also sobered many scientists beyond those who had been directly involved with it, pushing them to form and join organizations dedicated to attempting to control or abolish nuclear weapons. The antinuclear stance was a revival of the antiwar spirit that had enveloped the scientific community in the wake of World War I, and as deeply felt a revulsion at what science had wrought in the name of patriotism.

Today, when scientific machinery has become central to modern warfare, many scientists have similar doubts about participating in efforts that have military implications, if not direct applications. The categories of offensive and defensive weapons are cited again in the headlines about the "missile-defense shield," bringing to mind the doubts that some scientists raised during World War II as to whether they had misused their talents and expertise in their work for the government. While some scientists had readily worked on offensive weaponry, others had tried hard not to do so, confining themselves to such tasks as measuring the speed of sound in water, which they did not have to connect directly in their minds to the ability of a torpedo to tear a hole in the side of a ship, killing people belowdecks. As the war progressed, most scientists eventually accepted that killing enemy soldiers (and even civilians) was a necessary step in defeating the forces of darkness and achieving the climate of freedom that they believed fundamental to the continuation of democracy and to the full and open pursuit of scientific knowledge. The proximity fuze was originally intended as a defensive weapon designed to explode against an attacking enemy aircraft—but the technology could also be used, and later in the war was used, in a howitzer shell that would explode in proximity to massed troops on the ground, causing hundreds of casualties at once. Some scientists involved said later that had they known in advance of the proximity fuze's "offensive" use, they might have paused in designing it.

They discovered—as we must today—that the terms "offensive" and "defensive" are always arbitrary. Radar could (defensively) prevent enemy bombers from dropping their bombs only if the technology enabled friendly fighters to find the enemy's bombers early enough to (offensively) blast them from the sky. Was it for offensive or defensive purposes that a scientist helped develop a new poison gas in order to discover whether an enemy scientist could have similarly created the same or a similar gas? Could a scientist properly design a (defensive) gas mask unless he or she knew more about the (offensive) harmful gases it would have to filter out? Encountering a series of such questions, it became difficult for anyone engaged in militarily related scientific work to draw the line between the use and the abuse of science in general, and of their own science in particular; and what became even harder for a science worker to accomplish was creating a way of not crossing the line between the use of science and its abuse. Was it an abuse of science to develop an atomic bomb? An incendiary? A bullet? To learn what conditions in the sea would slow down or speed up sound traveling through liquid?

Regarded from a distance of more than fifty years, the distinction between the practical research done by scientists and technologists for the Allies and that done by similarly trained people for the Third Reich blurs. We must ask the same question of each country's scientific cadres: Did they do the bidding of their governments, without question or resistance? In doing so, did they temporarily compromise the purity of their science? Did ideology enter into and color their scientific judgment? The answer to all these questions, in all cases, was yes. World War II changed science permanently, moving it toward a relationship with the enveloping society in which the practice of science, whether basic or applied, is seldom apolitical or free from the influence of ideology.

The first wave of histories of World War II, published soon after the war's end, celebrated the generals, admirals, and political leaders of the democracies, ascribing to them the lion's share of the credit for winning the war. In Churchill's majestic five-volume history, which highlighted his own role, three pages in his first volume mention the development of radar and only one seventeen-page chapter in the second volume outlines what he called "The Wizard War," in which machinery with beyond-human sensing capacities was summoned, near-magically, to assist the hero commanders. The former prime minister would likely have been more generous with his prose on the subject of the scientific contributions had there not still been security restrictions on many of the matters.

Because the Cold War had begun when these books were published, many of them—Churchill's being something of an exception—minimized the contributions of the Soviet Union to the Allied victory in Europe; later historians argue persuasively that if Stalin's forces had not kept Hitler's occupied, the progress of the war and perhaps its outcome would have been different.

A second wave of books, cresting ten to fifteen years after the war ended, credited the millions of troops in the foxholes, backed by the massive production power of the United States, with winning the war. No one should underestimate the contributions of the fighting forces to the outcome of World War II. But the anonymous hero-tomes did give short shrift to the scientific and technological developments of the conflict; some stated openly that the good guys had won despite having weaponry inferior to that possessed by the enemy, and many of the books featured anecdotes relating how a piece of complicated equipment failed to work and had to be patched together or tossed aside. In one such book, the daredevil acrobatics and gallant sacrifices of the British pilots during the Battle of Britain took up most of the text and illustrations, while the equally decisive contributions to that victory of the Home Chain radar stations—and their mostly female personnel—merited only a few pages.

A third wave of historical evaluations of the war, published beginning in the late 1970s, revealed and emphasized the key role of Ultra and other intelligence decoding in providing the political leaders and generals with information that permitted resources to be most effectively applied against the enemy. In these books the military genius of such men as Eisenhower, MacArthur, and Halsey was undercut, tempered by revelations that their most celebrated strategic and tactical decisions were based on secret knowledge of the enemy's position, strength, and intentions. But here, too, the seemingly mundane contributions to the victory of antimalarial medication in the Pacific

Theater, proximity-fuze antiaircraft shells above Antwerp, new sonar technology, and Hedgehog mortars under the sea were viewed as peripheral.

A more balanced view of the Allied victories over the Axis nations in World War II must henceforth acknowledge that the contributions of science and technology to those victories were central and considerable, worthy of inclusion in the pantheon of decisive factors along with the sagacity of the democracies' political and military leaders, the tenacity of the Soviets, the determination, skill, and courage of the Allied nations' troops, the might of American manufacturing, and the advantages provided by code breaking and other intelligence.

Notes

Documentation of World War II is voluminous, although the portion of it pertaining to the scientific war is a bit slimmer. Any inquiry begins, in the United States, with the collections of the Library of Congress and the National Archives, which contain many records of the Third Reich in addition to those that touch on American science and technology efforts during the war, and with the collection of interviews conducted for the American Institute of Physics. In Great Britain the Imperial War Museum and the British Library have similarly large collections. Soviet and Japanese science and technology records are scarcer; historians hope that the new openness in Russia and a renewed willingness in Japan to reexamine the past will reveal more details and lead to greater understanding of their science communities' roles in the war.

New information regarding scientific efforts in the war continues to emerge in the United States, Great Britain, and Germany; as examples, the first biography of Vannevar Bush, by Zachary, was published in 1997, an extensive history of the development of radar, by Buderi, in 1996, and the most factual compendium about science in Nazi Germany, edited by Renneberg and Walker, in 1993.

In the notes that follow, designed to be read in conjunction with the bibliography, I have listed sources that are the most accessible to readers seeking further information about the contributions of science and technology to the conduct of the war. The notes are cumulative: sources cited in early chapters have been used also in later ones and so are generally not mentioned twice.

TERROR FROM THE SKY

The clash of V-1s and their robotic defenders is reported in many accounts, such as Macksey, *Technology and War*. Among the more unusual is Bates, *Flying Bombs Over England,* which was written during the war but whose publication was delayed fifty years. The role of the proximity fuze is chronicled by Baldwin, *The Deadly Fuze*.

PROLOGUE: LEGACIES OF THE GREAT WAR

Monthly issues of scientific magazines such as *Nature,* for the United Kingdom, and *Science,* for the United States, have been used to track attitudes of scientific and political establishments here and through the book. Another source used throughout is the multivolume *Dictionary of Scientific Biography,* for information about individual scientists such as Moseley, Haber, and the Braggs. This chapter draws extensively on the thoughtful analyses of Zuckerman and Medawar and on biographies of Churchill, Roosevelt, Hitler, and Stalin. Alfred Rosenberg is quoted in Andreas Heinemann-Grüder's chapter in Renneberg and Walker.

CHAPTER ONE: JOURNEYS TOWARD CONFLICT

Ipatieff's autobiography and the histories of Medvedev and Vucinich are good sources for information about Soviet science prior to World War II. Standard English-language books on Japanese science, such as Sigurdson, tend to focus on postwar developments; glimpses into Japan's prewar science can be gleaned from articles by Ichikawa, Shimao, and Yamazaki. The emperor's scientific background is delineated in Wetzler, and books by S. Harris, Bryden, and Tanaka detail the excesses of Major Ishii.

Recent and exhaustive biographies of Vannevar Bush and James Conant, along with their autobiographies, depict the activities of these pioneers. Frederick Lindemann has not been particularly well served by his biographers, who—as his patron Churchill does in the prime minister's memoir—laud Cherwell's positive influence without being specific and minimize his excesses. A more balanced assessment of Tizard is made possible by Clark's excellent biography and his *Rise of the Boffins*. Goldsmith's biography of Joliot, fulsome but marred by his admiration for leftist politics, is balanced by Weart's more scholarly *Scientists in Power*.

Used throughout this book are the first-rate studies on German science by Beyerchen (physics), Deichmann (biology), Macrakis (the Kaiser Wilhelm Institutes), and Walker (nuclear science), as well as individual biographies of Lise Meitner and Werner Heisenberg.

CHAPTER TWO: GERMANY, 1933–34

Wernher von Braun's history is recounted in Bower and in Piskiewicz. The Beyerchen, Deichmann, Macrakis, and Walker studies of German science provide important details on the effects of the 1933 decrees on individual scientists and the scientific community, as does Goran's biography of Fritz Haber. Benno Müller-Hill's introduction to Deichmann is particularly insightful. Borkin's classic study of the I.G. Farben conglomerate provides more details on Carl Bosch.

Early studies of refugee scientists by Hartshorne and by Bentwich have been supplemented by the later and more nuanced work in the books edited by Ash and Sollner, and Jackman and Borden.

Bloch's recollections are in his 1968 oral history interview for the American Institute of Physics (AIP). Ewald's recollection of the circumstances surrounding Planck's "Heil Hitler" comes from Ewald's 1962 AIP oral-history interview.

CHAPTER THREE: UNTIL THE SHOOTING BEGAN

The second volume of Gilbert's *A History of the Twentieth Century,* covering mid-1933 to 1951, provides many benchmarks for this era. Particularly useful for the early history of radar are the individual country chapters in Burns's compilation, the first volume of Guerlac, Buderi, and Sapolsky. Tuve's recollection is from his 1967 AIP oral history interview. Work on codes is documented in Hinsley's masterly three-volume study. Hinsley and Stripp's compilation, and Smith, trace the contributions of Bletchley Park.

Documentation for Soviet efforts, cited above, is supplemented by the studies of Kneen and of Parrott, who quotes the bragging Soviet scientist. Lysenko is quoted by Medvedev. The section on Kapitsa and Landau relies on my *Absolute Zero and the Conquest of Cold* (1999). New details of Rutherford's mediation of the Kapitsa affair were recently provided to me by Sir Brian Pippard.

Far-reaching books by Avery and by Bryden detail Canada's work on chemical warfare; Bryden quotes the Abyssinian general. The effort to keep the poison-gases island secret is documented in Cook and Cook. The American chemistry text cited is by Hessel, et al.

Technological advances, such as those tested at the underground facility, are documented in Ford, *German Secret Weapons*. Adolph Buseman's observation is from his 1979 AIP interview. For the nuclear effort, information comes from a wide variety of sources, including the several books about Heisenberg, Meitner, and other principal researchers, as well as from Rhodes's exhaustive *Making of the Atomic Bomb*, and from AIP oral history interviews with Genter (1971) and Kowarski (1971). Further details on some of the German physicists come from Goudsmit's memoir.

The meeting of Mayer and Turner is recounted from Jones's interviews with the participants in *Reflections on Intelligence,* while details of Jones's own involvement come from his *Most Secret War*.

CHAPTER FOUR: WAR AND PHONY WAR

Many references here come from my book *The Phony War, 1939–1940.* The Raeder and Dönitz quotes, as well as details of the onset of the submarine war, can be found in the classic account of the German Navy by Bekker. Mrs. Churchill's comment about German promptness is cited by her husband. Early British mishaps and triumphs are found in Roskill (the British counterpart to Bekker), and the volumes by Jones and by Terrell. The conversation between Sachs and Roosevelt is repeated in Rhodes.

Neufeld's comments are from his chapter in Renneberg and Walker, adapted from his 1995 book. Speer's comments on the rocketeers come from his autobiography. The unexpurgated version of the Oslo Report is published in Jones's books. Details on some of the far-out British science projects emerge from Clark's *Rise of the Boffins,* from Terrell, and from Crowther and Whiddington. The assertion that only British nationals could work on radar is from H. G. Kuhn's 1984 AIP interview.

Roosevelt's dismay at the unfitness of draftees is documented in Cowdrey's book on military medicine. Goodeve's comment is cited in Terrell and other sources. Konrad Lorenz's actions have been the subject of intense scrutiny by Deichmann and others, who quote Lorenz's writings to buttress conclusions about his collaboration with the Nazis. The enforcement of the "mercy killing" laws is documented in Sereny, Lerner, and other books about the inception of the holocaust. The firming of the resolve of Western scientists emerges from the pages of *Nature* and *Science.*

Jacques Allier's mission to Norway has been best put in context in Dahl's recent study; Rhodes and Clark feature extensive quotes from the Frisch-Peierls memo and the cryptic telegram sent by Lise Meitner. Peierls's own recollections are in his 1969 AIP interviews. The coming together of the American scientific management team emerges from autobiographies and biographies of the principals and is also analyzed in Dupree's article, "The Great Instauration of 1940." Sherwood's characterization of Bush is from his classic study *Roosevelt and Hopkins: An Intimate Memoir.*

CHAPTER FIVE: BATTLES ABOVE BRITAIN

A spate of books published around the fiftieth anniversary of the Battle of Britain provide excellent reviews of extant literature and recently uncovered facts, among them Bowyer and Ray. Galland's autobiography, *The First and the Last,* is a good reflection of German perceptions of the air battle.

The Tizard Mission to the United States, one of the turning points in the scientists' war, is the subject of one of the best recent books on the war, David Zimmerman's *Top Secret Exchange.* Buderi picks up the story of ensconcing radar at MIT, and the role of radar waves in the battle above Great Britain; further detail on the latter is provided in Guerlac, Burns, and the memoir of Taffy Bowen. Alvarez's comments are from his autobiography. Tuve's are from his AIP interview. Joliot's story emerges from Weart and various AIP interviews.

The birth and utility of Operational Research is described briefly in Waddington, Crowther and Whiddington, and tangentially in Zuckerman and in Goldsmith's biography of Bernal. It is a field whose origins and impact still await a definitive study.

CHAPTER SIX: TOWARD PEARL HARBOR

Conant's work and the initial formation of OSRD are covered in his autobiography, the fine biography of Conant by Hershberg, in Bush's memoirs, and in the Zachary biography; Bush's fight with Admiral Bowen is further detailed in Sapolsky. The various contributions of individuals at MIT, such as Charles Stark Draper and Harold Edgerton, emerge from Burchard and from the memorial volume on Draper edited by Lees. The Beaverbrook broadcast is cited in Josephson's *Totalitarian Science.*

Canada's gearing up for war is chronicled by Avery and by biographies of Banting. The breaking of the Japanese "purple" code and the consequences of that break are covered by Drea, by Lewin, and in other books about the year leading up to Pearl Harbor.

CHAPTER SEVEN: 1942, YEAR OF TRIALS

The medical-science aspect of the Pacific war is one subject of the 1948 book edited by Andrus and the more recent Cowdrey. The Japanese Technology Board's activities are chronicled by Yamazaki. The Gentner-Joliot exchange is recounted in Gentner's AIP interviews.

Speer's autobiography is a major source for information about science in Germany after he took over from Todt; his stories, however, must be evaluated in the light of Sereny's critique. Runge's memoir of German radar efforts is a chapter in Burns. Experiments on guayule are reported in the pages of *Science* and in Irvin Ashkenazy, "As the Guayule Ball Bounces," *Westways*, September, 1977. The development of the proximity fuze and the character of Merle Tuve are documented by Baldwin. The panoply of American research efforts comes from the series of postwar books on the OSRD, such as Baxter, Noyes, and Thiesmeyer, and from Alvarez's autobiography. The possibility of a "gas Pearl Harbor" is examined in Avery and in Bryden. Archibald MacLeish's speech was respectfully quoted in *Nature.*

The controversy over Heisenberg's role and the quality of the threat to the Allies posed by the German nuclear effort continues to reverberate. Rose's book, the most recent study, is the source of the doubts raised here about the efficacy of the German effort. Hellmuth Trischler's study of resistance in the German scientific community is a chapter in Renneberg and Walker. Flerov's alarum call to Stalin is recounted in Holloway.

A. J. P. Taylor's lectures on the sweep of the war provide a backdrop for discussion of the Pacific war, and Costello summons many of its details. The scientific war in the Pacific is the purview of a volume edited by Roy MacLeod, and the general course of that war in 1942 is the subject of Mullins. Beardsley's article provides details on the scientific exchange (and lack thereof) between the democratic Allies and the USSR.

CHAPTER EIGHT: SEAGOING SCIENCE

This chapter draws on many sources to supplement the standard histories of the Battle of the Atlantic by Bekker and Roskill. Conferences in connection with the fiftieth anniversary of the close of that battle, held in 1995, such as the compilation by Howarth and Law and a book by Syrett, provide much new and refined information. Gardner's reevaluation of the role played by Ultra is also essential revisionist reading. The sources cited above for Operations Research detail the various contributions of that discipline to submarine and surface warfare.

Gannon traces the American progress on torpedoes, and Buell's biography of Admiral King puts the best face on the admiral's initial stiff-arm and later embrace of science and technology as an aid to naval warfare. The antics surrounding Geoffrey Pyke's floating-ice aircraft landing field are lovingly detailed by Avery. Dönitz's speech is in Bekker.

CHAPTER NINE: THE GREAT SHIFT

Dornberger's speech and other material about the German rocket program are in Piskiewicz. The articles in the Swiss medical review *Praxis* and in *Chemiker Zeitung* were reported in *Nature,* as were the ravages of diseases in German-controlled Europe.

The April 2, 1943, German conference on nuclear physics is covered by Goudsmit, as well as by the books, cited above, concerning Heisenberg and the nuclear program. The work of Blome, Schumann, and others in biology is chronicled in Deichmann.

The work of American psychologists is documented in Bray. The story of the growth of Los Alamos and its atomic bombs is detailed by Rhodes, and by Conant in his autobiography. Bloch's story is from his AIP interview.

"Grocer," "Mandrel," "Airborne Cigar," and other radar countermeasures are featured in articles in Burns. The story of chaff emerges from Churchill, Jones, Baxter, Buderi, and Clark's *Rise of the Boffins*. The Air Ministry memo about religious leaders' opposition to the bombing is quoted in books about Cherwell.

The extent of the American R&D effort was the subject of articles in *Science,* as was the progress of the Kilgore-Patman bill. The research behind the sabotage device called the "Pencil" is reported in Connor and that behind penicillin in

Libenau, in Andrus, and in Cowdrey, which also covers the vaccine against typhus. Kaempffert's article on the wonders that science would produce once the war ended is from the *New York Times,* March 28, 1943.

CHAPTER TEN: VENGEANCE WEAPONS

Piskiewicz, Neufeld, Baldwin, Bates, Jones, Speer, Lees, and Macksey provide various details about the V-1 and V-2 attacks and the British and American attempts to destroy them on their launching pads and build defenses against them. Eisenhower's observation on the importance of knocking out the V-1 sites is from his memoir. The Quebec conference is best reported by Avery, and Bohr's escape to Great Britain is in Rhodes.

Cherwell's secret memo to Churchill about an anthrax weapon is in Avery and in Harris and Paxman, as is the full story of the mustard-gas explosion at Bari and Eisenhower's dissembling about it. These books also discuss Churchill's apparent willingness to use poison gas in reaction to the V-1 attacks, while Jones provides an alternate explanation of the prime minister's intentions.

CHAPTER ELEVEN: ENDING THE WAR

The ALSOS mission is detailed in memoirs by its leaders, Pash and Goudsmit, in the various books about Joliot and about the German nuclear program, and in both Bower and Hunt, who recount the origins and consequences of the Paperclip operations. Bower quotes the memo about the discoveries at the Walterwerke.

The Japanese balloon bombs are detailed in Cook and Cook. New American-developed air weapons introduced late in the war are described in Boyce, Baxter and other books on the NDRC efforts. The story of a pigeon-guided missle is in Bray. Roosevelt's unequivocal rejection of gas warfare is in Harris and Paxman, and also in Avery. The Japanese reaction to Roosevelt's death is cited in Gilbert. Rhodes, Conant, Bush, Zachary, and other books detail the final preparations for the testing of the atomic bomb and the dropping of the bombs on Hiroshima and Nagasaki.

EPILOGUE: SCIENCE, TECHNOLOGY, AND THE POSTWAR WORLD

The Farm Hall transcripts have recently been printed in their entirety, after too many years of being kept secret, and are discussed cogently in Walker. Meitner's letter to Hahn is reprinted in Deichmann and discussed in Rife. The United Nations report on the reassimilation of Japanese science is featured in Ichikawa. W. A. Noyes's provocative summary statement about the contributions of American science to the war is in his introduction to the book he edited about the work of the chemistry section of the OSRD.

Selected Bibliography

Alvarez, Luis. *Adventures of a Physicist.* New York: Basic Books, 1987.

Andrus, E. Cowles, et al., eds. *Advances in Military Medicine Made by American Investigators Working Under the Sponsorship of the Committee on Medical Research.* Boston: Little, Brown, 1948.

Ash, Mitchell G., and Alfons Sollner, eds. *Forced Migration and Scientific Change: Émigré German-Speaking Scientists and Scholars after 1933.* Washington, D.C.: German Historical Institute, Cambridge University Press, 1996.

Avery, Donald H. *The Science of War: Canadian Scientists and Allied Military Technology During the Second World War.* Toronto: University of Toronto Press, 1996.

Baldwin, Ralph B. *The Deadly Fuze: The Secret Weapon of World War II.* San Rafael, Calif.: Presidio Press, 1980.

Bates, H. E. *Flying Bombs Over England.* Kent, England: Froglets Publications, 1995.

Baxter, James Phinney III. *Scientists Against Time.* Boston: Little, Brown, 1946.

Beardsley, E. H. "Secrets Between Friends: Applied Science Exchange Between the Western Allies and the Soviet Union During World War II." *Social Studies of Science,* vol. 7 (1977).

Bekker, Cajus. *Hitler's Naval War.* Garden City, N.Y.: Doubleday & Co., 1974.

Bentwich, Norman. *The Rescue and Achievements of German Scholars.* The Hague: Martinus Nijhoff, 1953.

Béon, Yves. *Planet Dora: A Memoir of the Holocaust and the Birth of the Space Age.* Boulder, Colo.: Westview Press, 1997.

Beyerchen, Alan D. *Scientists Under Hitler: Politics and the Physics Community in the Third Reich.* New Haven: Yale University Press, 1977.

Birkenhead, Frederick E. S. *The Prof in Two Worlds: The Official Life of Professor F. A. Lindemann, Viscount Cherwell.* London: Collins, 1961.

Borkin, Joseph. *The Crime and Punishment of I. G. Farben.* New York: Free Press, 1978.

Bowen, E. G. *Radar Days.* Bristol, England: Adam Hilger, 1987.

Bower, Tom. *The Paperclip Conspiracy.* London: Michael Joseph, 1987.

Bowyer, Michael J. F. *The Battle of Britain 50 Years On.* Wellingborough, England: Patrick Stephens Limited, 1990.

Boyce, Joseph C., ed. *New Weapons for Air Warfare.* Boston: Little, Brown, 1947.

Bray, Charles W. *Psychology and Military Proficiency, A History of the Applied Psychology Panel of the NDRC.* New York: Greenwood Press, 1969.

Bryden, John. *Deadly Allies: Canada's Secret War.* Toronto: McClelland & Steward, 1989.

Buderi, Robert. *The Invention That Changed the World.* New York: Simon & Schuster, 1996.

Buell, Thomas B. *Master of Sea Power: A Biography of Fleet Admiral Ernest J. King.* Boston: Little, Brown, 1980.

Burchard, John. *Q.E.D.: M.I.T. in World War II.* New York: John Wiley & Sons, 1948.

Burns, Russell, ed. *Radar Development to 1945.* London: Peter Peregrinus, Ltd., and the Institution of Electrical Engineers, 1988.

Bush, Vannevar. *Endless Horizons.* New York: William Morrow, 1946.

———. *Modern Arms and Free Men: A Discussion of the Role of Science in Preserving Democracy.* New York: William Morrow, 1949.

———. *Pieces of the Action.* New York: William Morrow, 1970.

Churchill, Winston. *Their Finest Hour, The Second World War, Volume II.* Boston: Houghton Mifflin, 1949.

Clark, Ronald W. *The Rise of the Boffins.* London: Phoenix House, 1962.

———. *Tizard.* Cambridge: MIT Press, 1965.

Conant, James B. *My Several Lives: Memoirs of a Social Inventor.* New York: Harper & Row, 1970.

Cook, Haruko Taya, and Theodore F. Cook. *Japan at War.* New York: New Press, 1992.

Cowdrey, Albert E. *Fighting for Life: American Military Medicine in World War II.* New York: Free Press, 1994.

Crowther, J. G., and R. Whiddington. *Science at War.* New York: Philosophical Library, 1948.

Dahl, Per F. *Heavy Water and the Wartime Race for Nuclear Energy.* Bristol, England: Institute of Physics Publishing, 1999.

Deichmann, Ute. *Biologists Under Hitler.* Cambridge: Harvard University Press, 1996.

Deighton, Len. *Blood, Tears and Folly: An Objective Look at World War II.* London: Jonathan Cape, 1993.

DeVorkin, David H. *Science with a Vengeance.* New York: Springer-Verlag, 1992.

Drea, Edward J. *MacArthur's Ultra: Code-breaking and the War Against Japan, 1942–1945.* Leavenworth, Kansas: University Press of Kansas, 1992.

Dupree, A. Hunter. "The Great Instauration of 1940: The Organization of Scientific Research for War," in Gerald Holton, ed., *The Twentieth-Century Sciences.* New York: Norton, 1972.

Eisenhower, Dwight D. *Crusade in Europe.* New York: Doubleday, 1948.

Ford, Brian J. *German Secret Weapons: Blueprint for Mars.* New York: Ballantine Books, 1969.

Forman, Paul, and José M. Sánchez-Ron, eds. *National Military Establishments and the Advancement of Science and Technology.* Dordrecht, Netherlands: Kluwer Academic Publishers, 1996.

Galland, Adolf. *The First and the Last.* London: Methuen, 1955.

Gannon, Robert. *Hellions of the Deep: The Development of American Torpedos in World War II.* University Park: The Pennsylvania State University Press, 1996.

Gardner, W. J. R. *Decoding History: The Battle of the Atlantic and Ultra.* Houndmills, England: Macmillan, 1999.

Gilbert, Martin. *A History of the Twentieth Century,* vols. 1 and 2. New York: William Morrow, 1997.

Goldsmith, Maurice. *Frédéric Joliot-Curie.* London: Lawrence & Wishart, 1976.

———. *Sage: A Life of J. D. Bernal.* London: Hutchinson, 1980.

Goran, Morris. *The Story of Fritz Haber.* Norman: University of Oklahoma Press, 1967.

Goudsmit, Samuel A. *ALSOS.* New York: Henry Schumann, 1947.

Guerlac, Henry E. *Radar in World War II.* New York: Tomash, American Institute of Physics, 1987.

Harris, Robert, and Jeremy Paxman. *A Higher Form of Killing: The Secret Story of Gas and Germ Warfare.* London: Chatto & Windus, 1982.

Harris, Sheldon H. *Japanese Biological Warfare, 1932–1945, and the American Cover-Up.* London: Routledge, 1994.

Hartcup, Guy. *The Effect of Science on the Second World War.* Houndmills, England: Macmillan, 2000.

Hartshorne, Edward Y. *German Universities and National Socialism.* London: Allen & Unwin, 1937.

Hershberg, James G. *James B. Conant: Harvard to Hiroshima and the Making of the Nuclear Age.* New York: Knopf, 1993.

Hessel, F. A., et al. *Chemistry in Warfare: Its Strategic Importance.* New York: Hastings House, 1942.

Hideomi, Tuge, ed. *Historical Development of Science and Technology in Japan.* Tokyo: Japan Cultural Society, 1968.

Hinsley, F. H. *British Intelligence in the Second World War,* vols. 1–3. Cambridge: Cambridge University Press, 1979–83.

——, and Alan Stripp, eds. *Codebreakers, The Inside Story of Bletchley Park.* Oxford: Oxford University Press, 1993.

Holloway, David. *Stalin and the Bomb.* New Haven: Yale University Press, 1994.

Hooks, Gregory. *Forging the Military-Industrial Complex: World War II's Battle of the Potomac.* Urbana: University of Illinois Press, 1991.

Howarth, Stephen, and Derek Law, eds. *The Battle of the Atlantic, 1939–1945.* Fiftieth Anniversary International Naval Conference. Annapolis, Md.: Naval Institute Press, 1994.

Hunt, Linda. *Secret Agenda: The United States Government, Nazi Scientists, and Project Paperclip, 1945 to 1990.* New York: St. Martin's Press, 1991.

Ichikawa, Hiroshi. "Technological Transformation of Occupied Japan," *Historia Scientiarum,* vol. 5-2 (1995).

Jackman, Jarrell C., and Carla M. Borden, eds. *The Muses Flee Hitler: Cultural Transfer and Adaptation, 1930–1945.* Washington, D.C.: Smithsonian Institution Press, 1983.

Jones, R. V. *Most Secret War.* London: Hamish Hamilton, 1978.

——. *Reflections on Intelligence.* London: Heinemann, 1989.

Josephson, Paul R. *Totalitarian Science and Technology.* New Jersey: Humanity Press, 1996.

——. *Red Atom: Russia's Nuclear Power Program from Stalin to Today.* New York: W. H. Freeman, 2000.

Kevles, Daniel J. *The Physicists.* New York: Knopf, 1978.

Kneen, Peter. *Soviet Scientists and the State.* London: Macmillan, 1984.

Larrabee, Eric. *Commander-in-Chief: F. D. R., His Lieutenants, and Their War.* New York: Harper & Row, 1987.

Lees, Sydney, ed. *Air, Space and Instruments: Charles Stark Draper Anniversary Volume.* New York: McGraw-Hill, 1963.

Lerner, Richard M. *Final Solutions: Biology, Prejudice, and Genocide.* University Park: Pennsylvania State University Press, 1992.

Lewin, Ronald. *The Other Ultra.* London: Hutchinson & Co., 1982.

Libenau, Jonathan. "The British Success with Penicillin," *Social Studies of Science,* vol. 17 (1987).

Macksey, Kenneth. *Technology and War.* London: Arms and Armour Press, 1986.

MacLeod, Roy. "The Chemists Go to War: The Mobilization of Civilian Chemists and the British War Effort, 1914–1918," *Annals of Science* 50 (1993).

—— (ed.). *Science and the Pacific War.* Boston: Kluwer, 2000.

Macrakis, Kirstie. *Surviving the Swastika: Scientific Research in Nazi Germany.* Oxford: Oxford University Press, 1993.

Medvedev, Zhores A. *The Rise and Fall of T. D. Lysenko.* New York: Columbia University Press, 1969.

Moreno, Jonathan D. *Undue Risk: Secret State Experiments on Humans.* New York: W. H. Freeman, 1999.

Noyes, W. A., Jr., ed. *Chemistry: Science in World War II Series.* Boston: Little, Brown, 1948.

Overy, Richard. *Why the Allies Won.* London: Jonathan Cape, 1995.

Owens, Larry. "Vannevar Bush and the Differential Analyzer: The Text and Context of an Early Computer," *Technology and Culture,* vol. 27, no. 1 (January, 1986).

Parrott, Bruce. *Politics and Technology in the Soviet Union.* Cambridge. MIT Press, 1983.

Pash, Boris T. *The ALSOS Mission.* New York: Award House, 1969.

Perrett, Geoffrey. *A Country Made by War: From the Revolution to Vietnam—The Story of America's Rise to Power.* New York: Random House, 1989.

Piskiewicz, Dennis. *The Nazi Rocketeers: Dreams of Space and Crimes of War.* Westport, Conn.: Praeger, 1995.

Powers, Thomas. *Heisenberg's War: The Secret History of the German Bomb.* New York: Knopf, 1993.

Proctor, Robert. *Racial Hygiene.* Cambridge: Harvard University Press, 1988.

Ray, John. *The Battle of Britain: New Perspectives.* London: Arms and Armour Press, 1994.

Reingold, Nathan. *Science, American Style.* New Brunswick, N.J.: Rutgers University Press, 1991.

Renneberg, Monika, and Mark Walker, eds. *Science, Technology, and National Socialism.* Cambridge: Cambridge University Press, 1994.

Rhodes, Richard. *The Making of the Atomic Bomb.* New York: Simon & Schuster, 1986.

Richards, Pamela Spence. *Scientific Information in Wartime: The Allied-German Rivalry, 1939–1945.* Westport, Conn.: Greenwood Press, 1994.

Rife, Patricia. *Lise Meitner and the Dawn of the Nuclear Age.* Boston: Birkhäuser, 1999.

Rose, Paul Lawrence. *Heisenberg and the Nazi Atomic Bomb Project: A Study in German Culture.* Berkeley: University of California Press, 1998.

Roskill, S. W. *White Ensign: The British Navy at War, 1939–1945.* Annapolis, Md.: United States Naval Institute, 1960.

Sapolsky, Harvey M. *Science and the Navy: The History of the Office of Naval Research.* Princeton, N.J.: Princeton University Press, 1990.

Sereny, Gitta. *Into That Darkness: From Mercy Killing to Mass Murder.* London: Andre Deutsch, 1974.

Shachtman, Tom. *The Phony War, 1939–1940.* New York: Harper & Row, 1982.

Shimao, Eikoh. "Some Aspects of Japanese Science, 1868–1945," *Annals of Science* 46 (1989).

Sigurdson, Jon, and Alun M. Anderson. *Science and Technology in Japan*. London: Longman, 1991.

Smith, Michael. *Station X, The Codebreakers of Bletchley Park*. London: Channel 4 Books, 1998.

Speer, Albert. *His Battle with Truth*. London: Macmillan, 1995.

Syrett, David *The Defeat of the German U-Boats*. Columbia: University of South Carolina Press, 1994.

Tanaka, Yuki. *Hidden Horrors: Japanese War Crimes in World War II*. Boulder, Colo.: Westview Press, 1996.

Terrell, George. *Admiralty Brief: The Story of Inventions That Contributed to Victory in the Battle of the Atlantic*. London: George G. Harrap, 1958.

Thiesmeyer, Lincoln R., and John E. Burchard. *Combat Scientists*. Boston: Atlantic Little Brown, 1947.

Vucinich, Alexandr. *Empire of Knowledge: The Academy of Sciences of the U.S.S.R., 1917–1970*. Berkeley: University of California Press, 1984.

Waddington, D. H. *OR in World War 2: Operational Research Against the U-boat*. London: Elek Science, 1973.

Walker, Mark. *Nazi Science: Myth, Truth, and the German Atomic Bomb*. New York: Plenum Press, 1995.

Weart, Spencer R. *Scientists in Power*. Cambridge: Harvard University Press, 1979.

Weaver, Warren. *Scene of Change*. New York: Scribner's, 1970.

Weingart, Peter. "German Eugenics between Science and Politics," *Osiris*, vol. 5 (1989).

Wetzler, Peter. *Hirohito and War: Imperial Tradition and Military Decision Making in Prewar Japan*. Honolulu: University of Hawai'i Press, 1998.

Wilson, Thomas. *Churchill and The Prof*. London: Cassell, 1995.

Yamazaki, Masakatsu. "The Mobilization of Science and Technology During the Second World War in Japan," *Historia Scientiarium*, vol. 5-2 (1995).

Zachary, G. Pascal. *Endless Frontier: Vannevar Bush, Engineer of the American Century*. New York: Free Press, 1997.

Zimmerman, David. *Top Secret Exchange: The Tizard Mission and the Scientific War*. Montreal: McGill-Queen's University Press, 1996.

Zuckerman, Solly. *Scientists and War*. London: Hamish Hamilton, 1966.

Selected Oral History Interviews

DOZENS OF RECORDED and transcribed interviews of physicists from many nations, created for the American Institute of Physics' Center for History of Physics, are in its Niels Bohr Library in College Park, Maryland. Those interviews from which I have quoted directly are: Felix Bloch, conducted by Charles Weiner, 1968; Adolf Buseman, conducted by Steven Bardwell, 1979; P. P. Ewald, conducted by Thomas S. Kuhn and George E. Uhlenbeck, 1962; Wolfgang Gentner, conducted by Charles Weiner, 1971; Lew Kowarski, conducted by Charles Weiner, 1969–71; H. G. Kuhn, conducted by John Bennett and Anna Sheperd, 1984; R. E. Peierls, conducted by Charles Weiner, 1969; and Merle Tuve, conducted by A. B. Christman, 1967.

Acknowledgments

I WISH TO thank the following libraries in which research was conducted: the Library of Congress, the National Archives (I and II), the Franklin D. Roosevelt Library, the New York Public Library, the Alfred Lane Library at The Writers Room, the Niels Bohr Library at the American Institute of Physics, the Imperial War Museum, the British Library, the Scoville Library in Salisbury, and the libraries of New York University, Columbia University, Harvard University, and MIT. At all, the staffs have been knowledgeable, courteous, and helpful. Access to these depots, and to others not individually mentioned, through their on-line components has been revelatory as well as of assistance. For this project, as they have for other of my books in the past, the NYPL permitted me to use the facilities of its Wertheim Room, and The Writers Room provided sustenance well beyond the borders of its desks.

I wish also to express my appreciation to the following people who materially and spiritually assisted my progress: my wife, Harriet Shelare; my sons, Noah and Daniel Shachtman; Mel Berger; Jennifer Brehl; Donna Brodie; Russell Donnelly; Ralph Goldman; David Goodstein; Edward Nickerson; Edmond D. Pope; Stephen Power; Spencer Weart; Doron Weber; and many veterans of the scientific and technological front lines of World War II whose reminiscences and good will inform this work.

Photography Credits

PAGE 1

Photograph of physicists Samuel Goudsmit et al, courtesy AIP Emilio Segrè Visual Archives, Crane-Randall Collection.

Photograph of Goudsmit's team sifting through wreckage, courtesy of AIP Emilio Segrè Visual Archives.

PAGE 2

Photograph of Irène and Frédéric Joliot-Curie, courtesy of Société Francaise de Physiqe, Paris, courtesy of AIP Emilio Segrè Visual Archives.

Photograph of Lise Meitner & Otto Hahn, "Otto Hahn, a Scientific Autobiography," Charles Scribner's Sons, New York, 1966, courtesy of AIP Emilio Segrè Visual Archives.

Photograph of Pyotr Kapitsa, courtesy of AIP Emilio Segrè Visual Archives.

PAGE 3

Photograph of Ernest Thomas Sinton Walton, et al., courtesy of U.K. Atomic Energy Authority, courtesy of AIP Emilio Segrè Visual Archives.

Photograph of Churchill and Lindemann on June 18, 1941, courtesy of the Trustees of the Imperial War Museum, London (ref# H10786).

Photograph of Henry Tizard, courtesy of the Trustees of the Imperial War Museum, London (ref#: HU 54992).

Photograph of the Allied ship, courtesy of the National Archives.

Photograph of the High Frequency Direction Finding radar, courtesy of the National Archives.

PAGE 4–5

Photograph of the ground-based air intercept radar, courtesy of the National Archives.

Photograph of the control set, courtesy of the National Archives.

Photograph of the distortions on the radar, courtesy of the National Archives.

PAGE 6

Photograph of Vannevar Bush, courtesy of the National Archives.

Photograph of sorting cards at the U.S. National Register of Scientific and Specialized Personnel, courtesy of the National Archives.

Photograph of Franklin D. Roosevelt, et al., courtesy of the National Archives.

PAGE 7

Photograph of Shin-ichiro Tomonaga, et al., courtesy of the University of Tsukuba, Tomonaga Memorial Room, courtesy AIP Emilio Segrè Visual Archives.

Photograph of Robert Emerson, et al., courtesy of the National Archives.

PAGE 8–9

Photograph of the explosion of the German submarine U-175, courtesy of the National Archives.

PAGE 10

Photograph of the poster warning against the dangers of typhus, courtesy of the National Archives.

Photograph of penicillin being fed a mixture of corn steep liquor, inorganic salts, and sugar, courtesy of the National Archives.

Photograph of penicillin in a large-batch "fermentation" production, courtesy of the National Archives.

PAGE 11–12

Photograph of a V-1 robot bomb, courtesy of the National Archives.
Photograph of the V-2 on its launching pad, courtesy of the National Archives.
Photograph of the long-barreled gun, courtesy of the National Archives.
Photograph of the launch site, courtesy of the National Archives.

PAGE 13

Photograph of Walter Dornberger, et al., courtesy of the National Archives.
Photograph of Japanese rubberized-paper balloon, courtesy of the National Archives.

PAGE 14

Photograph of "Bat" guided missile, courtesy of the National Archives.
Photograph of experimental German guided missile plane, courtesy of the National Archives.

PAGE 15

Photograph of the atomic bomb, courtesy of the National Archives.

PAGE 16

Photograph of the damage from the atomic bomb, courtesy of the National Archives.

Index